Foreword

A long time ago on the light not It was nothing - neither land, neither sky, neither sand, neither cold waves. There was only one impenetrable black abyss, Ginnungagap, to the north of which lay kingdom eternal mists Niflheim, a to south - kingdom eternal fire Muspelheim. Muspelheim was creepy country sizzling heat, a in Niflheim, against, icy cold and darkness prevailed. The world was in chaos, and so it went on for a long time. How long - no one can say, because the time and space of the Eddic myths do not has nothing to do with the abstract concepts of extension and duration, which we are accustomed to operating with you. Mythological space is not only finite, but discrete and not uniform; it breaks up into isolated pieces, which are or place some important developments, or place stay hero. That's why it is absolutely impossible to make a map of the world of Eddic myths, since the countries in them mentioned are not oriented relative to each other. By the way, from here there is also such an important point as the lack of intelligible ideas about the world supersensible, or otherworldly, for all the worlds of Scandinavian myths are equivalent and equally real. They are not opposed to the "here-and-now" world in any way, but the possibility in them penetrate determined solely perseverance hero.

Others words narrator not looks on the items from outside and not trying portray them as they really appear to him. He places himself in the midst of events, inside what is happening, and does not think of itself outside this single whole. Not separating itself from the object, it sorts things and events first of all by their parameter significance. Considerations of reliability or visibility play no role for him. Similar absence clear opposition subject object can name internal dot view on space.

BUT because the space Eddic myths deprived connectivity and crumbles in fragmentary husk, then the declarative primordial emptiness is not conceived outside of the concrete filling. The abyss of the world, as it turns out on the very first pages, is not at all world, since from the north it adjoins the country of darkness and cold, and from the south - the kingdom of fire. Therefore, creation turns out to be not a birth out of nothing, but a banal transformation of existing. FROM so same success can rip open old worthless dress and carve out from him new suit.

When the life-giving spring Görgelmir suddenly burst forth in the realm of fogs, the abyss of Ginnungagap was rushed by the waters of twelve mighty streams. And though the fierce frost Niflheim immediately turned water in ice, source continued beat not ceasing.

Blocks of ice grew by leaps and bounds, heaping on top of each other and climbing up, and when a monstrous ice sheet crawled close to the outskirts of Muspelheim, its fiery breath melted the age-old ice. Fireworks of hot sparks, splashed from the realm of fire, mixed with melt water and breathed life into it. And then from the abyss of Ginnungagap slowly rose gigantic figure, trampling severe foot fixed ice shell. It was the giant Ymir, the first living creature in the world. On the first day of creation (if consider the birth of Ymir the first day) a boy and a girl appeared under his arm, and one leg conceived With another six-headed giant son. So It was supposed Start cruel and insidious tribe giants Grimtursenov.

Ymir and his offspring needed food, but in the darkness, cold and chaos of the lifeless feeding in the desert was very problematic. Therefore, along with the progenitor giants from melting ice appeared gigantic cow Audumla, from udder which four rivers of milk flowed. Audumla grazed in the ice and licked the salt ice lumps. She worked so hard that by the end of the third day, a giant of the Storm stepped out of the block, the forefather of the three gods - Odin, Vili and Be. The brothers did not favor the imperious and cruel Ymir, but therefore they rebelled against the first of the giants and, after a long exhausting struggle, killed his.

And peace was so huge what blood, gushing from his ran, flooded the whole world. The giants and the cow Audumla disappeared without a trace in the raging elements, and only one of Ymir's grandchildren were lucky: he managed to build a boat on which and escaped with his wife. The brother gods set about rebuilding the world, for the eternal cold and darkness that reigned around, they didn't like it. From the body of Ymir they made the earth in the form of a flat disk and they placed her in the middle of a vast sea that was formed from his blood. From the skull of Ymir brothers made heavenly vault, from his bones built the mountains, from hair made trees, from the teeth - stones, and from the brain - clouds. In the middle of the world they built Midgard - the abode people (in translation, midgard means "middle courtyard"), and the outlying lands on the seashore given to the giants. To protect against the giants, they surrounded Midgard with a high wall, which made from the eyelids (or eyelashes) of Ymir. Each of the four corners of the firmament gods rolled up in the shape of a horn and planted in each horn according to the wind. From the hot sparks flying out of Muspelheim, they made stars and decorated the firmament with them. Some of the stars were fixed motionless, and some were allowed to circle the sky so that they couldlearn time.

True, other Eddic songs say that heavenly bodies existed and before, that's why Work gods reduced Total only to instructions those places, which them should have take.

The sun did not knowwhere is his home the stars didn't knowwhere they shine didn't know for a month relics his.

Internal dot vision on the space appears, in in particular in volume, what geography in Scandinavian myths does not exist apart from ethics. All good things are gathered in center peace, a evil doomed huddle on the his outskirts. Any subject automatically receives a quality rating depending on where it is located. In the middle of the world located Midgard, and the country Giants Jotunheim lies on the outskirts, that is reasonable suppose what outskirts peace - this is land. Between topics from others songs follows, whatthe outskirts of the world is nothing but the sea encircling the earth in a ring, at the bottom of which slumbers the monstrous world serpent Jormungandr, biting his own tail. But when the gods go to the land of the giants, every time they have to cross the sea straits. The periphery of the Scandinavian universe paradoxically turns out to be land and by sea simultaneously.

AT center peace too reigns flagrant confusion. Except Midgard, inhabited people, the chamber of the gods Asgard rises there, and the world tree, the ash tree Yggdrasil, pierces terrestrial disk in accuracy in the middle for his crown extends above everyone the world. AT later Christian interpretations attempt to elevate Asgard to heaven, but these pitiful "grimaces and jumps" can only cause a condescending smile, since the sky of the Eddic myths is no different from the earth. And although in Euclidean space the combination of three objects in the same place is absolutely impossible, this storytellersabsurdity is not embarrassing. Just the chamber of the gods, the abode of people and the sacred tree are not may be nowhere but middle peace.

The time of the Scandinavian myths is also fragmentary and rigidly tied to the event row. If nothing worthy of attention is happening in the world, then time stands still. place. It is simply not conceived as a fluid substance, not subject to influences. from outside: if between two events missing causal connection, arrange them on order resolutely impossible. Let's say absolutely dont clear, in which chronological sequences must be located visit thunderer Torah to to the giant Geirod, his duel with the world serpent Jormungandr and the battle with the stone giant Grungnir. More Togo, any narration immediately crumbles on the fragments living an independent life, and the character of this or that myth is almost always a static figure performing a memorized circus number. There is no development in it. For example, Magni, the son of Thor, is famous for pushing the leg of the defeated giant from the neck father. However, this was not his childish feat, but a feat in general. Magni is always a child and outside of his courageous act simply does not exist. On the other hand, the father of the gods One, apparently always old man.

Past, present and future also smoothly flow into each other and wonderful coexist side by side. This is unambiguously evidenced by the grammar of the Eddic myths, when forms past time at ease alternate With forms present or future. The gods do not live in time, where events can turn so or something like that, but in a kind of motionless eternity, where everything is painted as if by notes. From mere mortals are separated by an absolute epic distance, as one smart historian. In that distant era, everything was different and even time flowed differently. Coming death gods, designated grinding word "Ragnarok" set out volva-soothsayers as an event taking place here and now, but this is not at all does not contradict the fact that the catastrophe has yet to occur. Others words past and future presented themselves equally real, and moving along the time axis seemed as natural as, say, traveling from Asgard in Jotunheim.

A fairly detailed retelling of the Scandinavian myth of the creation of the world has not been undertaken. out of love for art (although the gloomy and majestic poetry of the northern sagas cannot, in our look, leave indifferent a person with a good literary taste), but only for Togo, to you, reader, could imbue confusing cosmogony ancient. Acts Scandinavian gods and heroes in pre-Christian era received call Eddic myths, because what they reached before our days in two literary monuments - "Younger Edda" and "Elder Edda". The author of the "Younger Edda" is considered Icelander Snorri sturluson, which the in first half XIII century collected together and systematized the myths that existed in the oral tradition. However, to call him the author possible with a certain stretch, because at that time such a concept simply did not exist. The authorship of the "Elder Edda" has not been established, just as the etymology of the word "Edda" is unknown; supposed, what it going on from farms Oddy, where Snorri brought up but long away not all scientists such the interpretation is satisfactory.

Cosmogonic myths about the birth of the world from chaos existed at different times among many peoples. Nearly all they permeated one and topics same motive: original chaos opposing elements (how rule fire and water) on will gods pretends to be in well-organized space, and disorder gives way to strict harmony. Often the creator departs from affairs, and then committed transition from mythological time to time historical. In other words, the world is born not in time, but together with time. If a apply to ancient layers folklore and mythological performances, show up amazing resemblance cosmogonic systems, created in different parts of the globe. Of course, there will be no detailed match, but the main the line will be highlighted quite clearly: a fierce confrontation between the polar forces, fierce clashes of gods and monsters, ordering of primordial chaos and tedious repetition all change. ancient egyptian or Hindu cultural traditions in this sense not at all not different from antique. We decided apply to Scandinavian legendsonly because they have an eerie stamp of pagan authenticity, which is not you will find, for example, in ancient Greek myths, which in the course of centuries-old cultural the polishings are pretty worn out and look, so to speak, a bit postmodern against the background of the Eddic songs. Icelandic scientist Sigurd Nordal So wrote about one from books "Younger Edda":

...

Gylvi's Vision is one of those timeless works that you can read child immediately after the primer and then again and again at all levels of development and knowledge and each once find new, and new, and new. This book simultaneously and transparent and difficult to understand, simple as a dove, and cunning as a snake, depending on how deep reader penetrates in her. For, although pagan outlook not fully revealed in her, in greater wholeness his not find neither in what friend work.

When in era Enlightenment triumphed natural science painting peace, the naive ideas of the ancients were crossed out. The universe has become a pattern divine harmony, eternal and unchanged space, living on strict mathematical laws. At the end of the 19th century, they even began to talk about the end of physics: they say that all fundamental issues have already received final resolution, therefore left only take a walk hand masters on polished before shine facade, to eliminate minor roughness. However very soon from inconspicuous cracks threw such smoke that the whole building of traditional physics was desperately feverish. From the past there was no trace of goodwill left. The cozy Victorian era was slowly fading into past, and on the shift classical science XIX century came new physics - paradoxical unusual and frightening. change occurred on the turn centuries not bad reflected in famous comic quatrains.

This world was shrouded in deep darkness.Let there be light! And here comes Newton. But Satan not for long waited revenge:
Came Einstein - and became all how before.

Of course, it would be absurd to draw a direct parallel between natural philosophical views ancient and achievements contemporary natural sciences. However pagan painting peace at all his naivety and artlessness profitable is different from motionless and boring space determinists. She is paradoxical exquisite and amazingly dynamic. By the way, the thinkers of later epochs always abundantly drawn from folklore. For example, one of the most profound and original minds of Hellas - Heraclitus the Dark (VI century BC), who said that you cannot enter the same river twice, once proclaimed: "You should know that the war is universal!" Of course, this is not about armed clashes on the field scolding, because they Total only private happening universal law: all being - fetus wrestling, and myself world there is eternal becoming.

Pagan natural philosophy is far from being as primitive as it might seem on the surface. first glance. Let's say myths about the beginning of time the universe was more in state, close to chaos, reveal surprising intersections with the latest cosmological ideas. True, the ratio of chaos and space, entropy and order in modern cosmological models of the birth of the Universe from nothing is somewhat different: the first moments the lives of our world are conceived as a state of a high order, and further entropy irresistibly is growing. However, exists and opposite dot vision: "primary atom", from whom arose world, was chaotic homogeneous state, a all story The Universe is nothing but the process of its structuring, evolutionary complication. One way or another, but the fundamental questions of being were again in the spotlight astrophysicists and cosmologists, of course, on the friend level understanding.

Modern physical painting peace lost visibility, former alpha and omega classical science before last centuries. When reading about quantization space, corpuscular-wave dualism or amazing metamorphosis, that occur over time inside black holes, one involuntarily recalls a split into pieces space Eddic myths and amazing mythical time, not knowing differences between past and future. And the combination in one point of the world of people, the hall gods and the sacred world tree - why not the frills of elementary particles in physics microworld? The miraculous evaporating of the Universe from the space-time foam and its inevitable death, when "time will be no more" (the words of John the Theologian), it is also possible find correspondences in the myths of different peoples. Therefore, it is hardly reasonable to pat ancestors on the shoulder, complaining about the limitations of their natural science knowledge. Not known yet, what is easier - to come up with a new cosmological scenario or be the first to give answers, let approximate or even erroneous, to questions about fundamental patterns being. And who knows, perhaps, sophisticated models of the world order, which are much modern astrophysics, will seem to our descendants as clumsy and distant from reality, what us see cosmogonic representation ancient.

distances, miles, miles

The one who created the world made a pipe dream of meeting Created on different stars. He erected between them a barrier perfect empty and invisible but irresistible: own, a nothuman distance.

Stanislav Lem

AT antiquity people lived on the flat Earth. Nothing amazing in this No, for human eye earthly surface and indeed sees running away per horizon boundless plane, if, certainly, neglect local drops relief on height. Traveling through the valleys and over the hills, the merchants and soldiers of the Ancient World could own experience to make sure that the surface of the Earth is a huge flat pancake.

However, to consider our distant ancestors as naive simpletons would be reckless and shortsighted. It's just that science at that time was still floundering in diapers. Loose pile facts, where accurate observations and amazing conjectures interspersed with monstrous misconceptions, yet to be systematized. Separating the wheat from the chaff is not at all such easy task like it could seem on the first glance.

But if vision us not deceives and Earth really flat, should would to find out, how long away she is extends. BUT because the nobody from mortals not managed get to its edge and look down, it seemed quite logical to assume that this the edges no at all - earthly surface nowhere not ends. But infinity - very uncomfortable concept, poorly amenable rational comprehension, and people always sought from her get rid of. If a same edge at Earth after all there is, what, tell on the mercy, can prevent the waters of the world, from all sides washing the land, without a trace to pour into bottomless an abyss? Position saved heavenly vault, knocked over above earth gigantic bowl and constituting with it a single whole. So forever on the run the horizon will be the place where the crystal dome of heaven connects with the earth's firmament. Between by the way, biblical expression " firmament earthly and firmament heavenly" is echo those Old Testament geographic representations.

So, we at the very least sorted out With device Universe. Happened trough With flat-bottomed, slammed with the lid of the vault of heaven. It remains to determine the form and dimensions this designs. However at different peoples sometimes existed diametrically opposite opinions on this account.

Say, the ancient Egyptians, who lived in the Nile Valley, and the Sumerians, who inhabited the interfluve Tigris and Euphrates, believed that the Earth is much longer from east to west than from the north South. For a number of historical reasons, they were fairly well acquainted with the inhabitantsneighboring countries lying at the eastern and western borders of their kingdoms, but the southern and northern land for a long time were for them nearly complete terra incognita. That's why Sumerians and Egyptians Earth drawn in form rectangular drawer, elongated in latitudinal direction. Among the Greeks, the sense of geometric proportions was, apparently, developed better: in their opinion, the Earth was a round plate, of course, with Greece in center. land co all parties washed water mighty rivers under name Ocean, a mediterranean sea was her skinny branch, his kind appendix, stretched out to the center peace.

ancient greek historian and geographer Hecataeus Milesian, lived for five centuries before the beginning of the Christian era, the author of the fundamental work "Earth Description", which came to to the present day in fragments, even tried to calculate the dimensions of this plate. He came to the conclusion that its diameter should not exceed 8 thousand kilometers; so the area flat earth will be equal to 50 million square kilometers. And although true the area of our planet is 10 times larger, we dare to believe that the figures obtained by the bravea native of Miletus, seemed monstrous to contemporaries. Of course, the circle is more perfect figure on comparison With clumsy rectangle, but the sacramental question of what holds the earth's disk in place still remained without response. The ancient Greeks were not born out of thin air and knew perfectly well that all heavy bodies have trend fall down.

– If the flat earth disk is really so large, the skeptics said, joyfullyrubbing dry palms, then let the respected Hecateus explain to us, unreasonable, what forces cause it to hang motionless. If, nevertheless, he falls with a whistle into emptiness, like all other bodies, then why do we not notice this impetuous fall?

We not we know how answered the first antique geographer on the uncomfortable questions opponents. It was easiest to say that the earth's firmament extends downward indefinitely, but it immediately led to memory cursed infinity, from which just now managed get off. Where smarter It was suppose what terrestrial disk rests on the anything durable. Hindus put the earth on the four pillar.

– Highly Good, - scathingly filtered through lip skeptics, - a on the how stand pillars?
– On the giant elephants, this is even small children know.
– BUT elephants?
– BUT elephants, Yes will be to you known trample on their feet shell giganticturtles.
– BUT turtle?...

Evil infinity stubbornly crawled out of all holes time after time, and the idea of flat Earth drove thinker in hopeless dead end.

Let's remember the funny tale of Lazar Lagin about the powerful genie Gassan Abdurrahman ibn Hottabe by birth from ancient Arabia, by will fate found himself in modern Moscow. They say he was a very influential figure in the court of the wise king Solomon, who ruled 3000 years ago in Palestine, but somehow did not please Caesar. The loving king (according to legend, Solomon had 700 wives and 300 concubines) did not to stand on ceremony with the disobedient and without long conversations ordered him to be imprisoned in an earthen vessel, which was to be drowned in the depths of the sea. And 3000 years later, a Moscow schoolboy Volka Kostylkov by chance came across on the mossy ceramic vessel in time morning bathing. How live genies, in accuracy nobody not knows but Hottabych turned out to be an unusually cheerful and accommodating old man, and therefore immediately offered his savior a lot of services. Volka had an exam in geography, in which he was rather finely swam So what after several purely formal body movements right pioneer and valid member astronomical Cup at Moscow planetariums waved mutually beneficial deal.

Hints genie - not lb. raisins. Wolke got India, but about Arabian sea and the Bay of Bengal, which wash the shores of this vast peninsula, poor boy did not have time to say anything. Against his own will, he spoke utter nonsense about country, lying on the himself edge earthly disk, and about adjacent lands, inhabited bald people which eat exclusively raw fish and woody cones.

When they asked him what disk he was talking about, and did he not know that the Earth has the shape of a ball, Volka, obeying Hottabych, grinned arrogantly and continued in toy same eloquent manner:

...

– You if you please tell jokes above yours most devoted student! If a would Earth was ball, water flowed down would With her way down and people died would from thirst a plants dried up. Earth, O most worthy and noblest of teachers and mentors, had and has the form flat disk and is washed on all sides by a majestic river called "Ocean". The earth rests on six elephants, and they stand on a huge tortoise. That's how the world works, oh teacher!

jokes jokes but narrow-minded representation about nature of things on the rarity tenacious. It is said that once Bertrand Russell, the eminent English philosopher and mathematician, gave a public lecture on astronomy. And although it happened relatively recently, at the beginning of the last century, the lecturer was thorough and unhurried. Talking about how The earth revolves around the sun, he did not fail to notice that our magnificent daylight light is ordinary star and, in my turn, too moving around center Galaxies. When the lecture came to an end, a small elderly woman rose from the back rows. lady and stated what all, about how here interpreted dear lecturer, - continuous nonsense.

– On the himself deed, - said she is, - our world - this is big flat plate, which costs on the back giant tortoise.

– Well Good, - smiled Russell, a on the how same holding on turtle?

– You are very perceptive, young man, answered the little old lady. - A turtle is standing on the back of another turtle, that one is on another, and so on, and so on, and so on. Further.

Maybe, cosmogony Hecatea more for a long time was would in go, if would not separate annoying little things. Observant Greeks noticed that the picture of the starry sky is palpably changes as you travel from south to north. Part of the stars floats over the southern horizon, and in the north, new constellations light up that cannot be seen in the southern latitudes. For example, Polar star step per step climbs all above and above, from what With it was necessary to conclude that sooner or later it would hang right overhead traveler. Of course Greeks It was unaware, what similar event maybe take place only only on the Northern pole, but trend spoke herself per myself. (In fairness, we note that five centuries before the birth of Christ, the Polar, that is alpha Ursa Minor, was not the star closest to the pole, but these particulars we are here omit.) On the other hand, when traveling south, the North Star begins to slide down, dragging along the northern constellations, and unfamiliar ones emerge from the southern horizon stars. On the equator line (the concept is equally speculative for the ancients Greeks, like North Pole) The North Star should lie on the northern horizon. If the earth were a flat disk, the pattern of the constellations would change extremely slightly, shifting slightly by perspective. The starry sky would look the same everywhere, but the complex evolution not would and in remember.

Therefore, the ancient Greek philosopher Anaximander, who lived almost 100 years before Hecateus and too native Miletus, suggested what earthly surface warps on direction from south to north. Instead of a round slab, he got a cylinder lying horizontally, on the surface of which people live. It must be said that the Asia Minor city Miletus was the real cultural Mecca of the ancient world, for an older contemporary Anaximander, his countryman and teacher Thales the first representative schools Ionian natural philosophers, too understood sense in movement heavenly luminaries. By legend, he predicted a solar eclipse of 585 BC. e. Frankly, it's not entirely clear how he managed to do this, because at Thales our Earth had the shape of a flat disk, floating on the surface of the endless ocean. The theory of solar and lunar eclipses the Greeks developed much later, so let's leave the achievements of Thales of Miletus to conscience chroniclers.

The cylindrical Earth of Anaximander was an undeniable step forward compared to flat Universe of Hecateus or Thales, but, alas, she did not save the situation. As is known, antique Greeks were maritime the people very early mastered and settled mediterranean coast on the everyone his throughout - from Gibraltar pillars on the west to the shores of Asia Minor in the east. Nimble sharp-nosed ships of brave sailors not only penetrated through the chain of straits into the Black Sea (the Greeks called it Euxine Pontom), but also went to the Atlantic, and in search of the legendary country of Thule, they reached British islands (expedition Pytheas). not without reason fabulist Aesop once compared their fellow tribesmen with frogs, stuck around their native swamp on all sides. ancient Greeks, whose whole life was closely connected with the sea, almost every single day had a chance see off fragile shells in distant swimming. Carefully watching per ships leaving the harbor, they more than once had the opportunity to make sure that the ship does not just melt "in the blue fog of the sea", but seems to disappear behind the slope of the hill along parts: first the hull is hidden from the eyes, then the sail, then the tops of the masts. To those who able think, remained do elementary mental an effort, to come to conclusion about sphericity Earth. More Togo, ships eluded under mountain absolutely equally outside dependencies from directions, in which they floated. Travel on the south gave exactly the same result as sailing east or west. Cylindrical Anaximander's model was unable to explain the uniform curvature of the Earth's surface along all directions, and therefore turned out to be untenable. The Greeks rightly judged that only surface ball not contradicts all sum accumulated antique science facts.

It is believed that the idea of the sphericity of the Earth was first expressed by a contemporary Socrates Philolaus of Tarentum. This happened in the second half of the 5th century BC. e. And great Aristotle, who lived about 100 years later, already knew for sure that the Earth is a sphere, and even added his own argument to the treasury of ancient astronomy. He guessed that cause lunar eclipses is discarded earth shadow, when our planet is between the moon and the sun. Moreover, the cross section of the earth's shadow on the disk The moon is always round, which can only happen if the earth has form ball. Be Earth flat disk, painting was would absolutely different. They say, what

Aristotle even tried calculate length equator our planets, taking per basis the difference in the position of the North Star in Greece and Egypt. He got the size approximately equal to 400,000 stages. If we translate ancient measures of length into the familiar to us metric system, then in one stage there will be about 200 meters. Anyway, most historians believe that this is exactly the case (the Attic stages numbered 185 meters, a Babylonian - 195 meters), although complete clarity in this question no. So or otherwise, but the diameter of the Earth, measured by Aristotle, turned out to be twice the modern values.

But Eratosthenes of Cyrene, who lived in the III century BC. e., got much more reliable result. From the calculations of Eratosthenes it followed that the circumference of the globe is (in converted to metric measures) 39,700 kilometers (modern calculations give almost 40 000 kilometers). The result of Eratosthenes managed to be slightly corrected only at the end of the 18th century. centuries, which cannot but alert the thoughtful researcher, since the tools which enjoyed Greek astronomer, were on the rarity primitive. He measured the height of the Sun above the horizon on June 21, on the day of the summer solstice, when noon the luminary rises highest in the sky. Measurements were taken on the same day two Egyptian cities - Syene (modern Aswan) and Alexandria, which is located on the 800 kilometers north. AT Siena vertically stuck in earth stick not gave shadows, from what follows, what Sun in that day stood exactly in zenith above Siena. BUT here in Alexandria, a short shadow was revealed, which corresponded to the position of the noon sun on the 7 s superfluous degrees south zenith.

Be Earth flat, Sun and in Siena and in Alexandria stood would in zenith at the same time, since the distance between these cities is relatively small. BUT as soon as it was possible to identify the difference in the length of the shadow, this means that the surface of the planet between the cities is curved, since the sticks in Syene and Alexandria turned out to be at an angle to each other to friend. A simple calculation shows that if a difference of 7 degrees corresponds to 800 kilometers then difference in 360 degrees (full turnover on circles) will give value near
40 000 kilometers. Clear, what if known length circles, not will be labor calculate diameter ball, his volume and square his surfaces. Diameter Earth is approximately 12,800 kilometers, and the area of a sphere with such a diameter will be equal about 500 millions square kilometers.

By the way, humanity is very lucky that the size of the Earth is not particularly large. Whether our planet is much larger, the view of the starry sky when moving a few hundreds kilometers practically not changed would, a ships managed would melt away in atmospheric haze before their hull disappeared over the horizon. Yes, and the border of the earth the shadow on the disk of the moon would look like a perfectly straight line in this case. Guess by eye insignificant curvature It was would resolutely impossible. Necessary believe, what and development astronomy would then go completely differently, and the idea of the sphericity of the planet arose much later.

If a would Universe exhausted earth, ancient Greeks allowed would basic question of cosmology even more than 2000 years ago. However, there was also the sky. Because the It was irrefutably proven what Earth It has spherical shape, should to reconsider traditional ideas about the firmament of heaven. Inverted bowl model passed into the archive, and its place was taken by a hollow sphere, covering the globe from all sides. Clear, what diameter such spheres must to be more diameter Earth. Whole question is how much more. In other words, how far is the sky? Common bike about the fact that it is a little higher than the eagle flies, no longer worked. What interesting things can see in the sky? In addition to actively traveling through the firmament of the Sun and the Moon, in the sky there are also fixed stars. More precisely, they are shifting all at once, as if heavenly the sphere carries them along, making a complete revolution around the Earth every 24 hours. But friend relatively friend stars motionless, a picture constellations always one and that same. And through a year, and after 10, and after 100 years, they can be found in exactly the same place. One gets the impression that the stars are pinned to the celestial sphere, which relentlessly spinning around the Earth.

However, the ancients loved to observe and were able to notice. They have long discovered that a large star family has its own fidgets who do not sit still, but rush about like mad, drawing complex loop-shaped zigzags throughout the year. sun and The moon, of course - they are too big to be considered stars. Well, and more such hurried exactly five - Mercury, Venus, Mars, Jupiter and Saturn. The Greeks began to call these eternal wanderers planets, what in translation means "wandering". It turned out, what at famous dexterity can even define relative distances between them.

Nearer Total to earth, no doubt was Moon, because in time solar eclipses sailed between earth and Sun. Distances before others planets can calculate from relative their speeds movement against the background of fixed stars. We know from experience that the closer an object is, the faster it moves. Bird high in the sky soars majestically and slowly a being low above earth, rushes by like swift gray lightning. So, the alignment of the ancient Greeks looked like this (as distance from Earth increases): Moon, Mercury, Venus, Sun, Mars, Jupiter and Saturn.

This is how the geocentric model arose, which is usually associated with the name of Claudius. Ptolemy, ancient Greek astronomer, who lived in I–II centuries n. e., creator fundamental treatise Almagest. AT center universe still rested Earth, and around it revolved in regular circles eight nested one in another sphere carrying the Moon, the Sun and the five planets known by that time. On the the eighth sphere contained the fixed stars. To explain a very complicated way, which the planets commit on the background stars, Ptolemy suggested what they in addition move in smaller circles linked to the corresponding sphere. These additional orbits got the name epicycles.

BUT it is forbidden whether calculate not relative, a absolute distance although would before some celestial bodies? Except for the semi-legendary Aristarchus of Samos, supposedly who built the heliocentric model one and a half thousand years before Copernicus, for the first time the outstanding astronomer of antiquity Hipparchus took care of measuring the distance to the moon, lived in the 2nd century BC. e., almost 300 years before Ptolemy. Recall that during the moon eclipses on the disk Moon observed circuit terrestrial shadows, which always (at any eclipses) is a circle. By the curvature of the edge of the earth's shadow, one can judge the size of its cross section compared to the size of the moon. If we assume that The sun is much farther from the earth than the moon, you can calculate how far from the earth the moon must be positioned so that the earth's shadow is reduced to observable size (we know the dimensions of the Earth). Hipparchus came to the conclusion that the distance to the moon is 30 times more earthly diameter if to accept diameter value our planets, found Eratosthenes (12 800 kilometers), then distance before Moon will be 384 000 kilometers.

it absolutely brilliant result: on modern estimates, average distance between the moon and earth is 384 400 kilometers, changing from 356 610 kilometers at perigee (point of minimum distance) to 406,700 kilometers at apogee (point maximum removal). And so I am ready to agree with the revisionists of the orthodox historical version who insist that measurements of this level of accuracy are not could to be fulfilled before era Renaissance. More Togo, even in XVII century similar accuracy was daunting task. Absolutely unclear, what way the ancient Greeks managed to accurately measure the angles between celestial bodies using those primitive tools at their disposal. I no longer talk about that for accurate astronomical observations, a clock with a second arrow, while the mechanical clock invented in Europe at the end of the Middle Ages long time not have had even minute. Between topics us tell, what Hipparchus With

with breathtaking accuracy calculated the duration of the lunar month - 29 days 12 hours 44 minutes 2.5 seconds (actual value - 29 days 12 hours 44 minutes 3.5 seconds). How he managed make a mistake Total on the one give me a sec (and how thought halves seconds), not having mechanical hours, story is silent.

Chronicles report what distances between geographic paragraphs Eratosthenes measured by the speed of camel caravans, and determined the angles of the Sun's rise using sticks dug into the ground. It looks like the truth, because, say, among the medieval Mongols, one length was considered a daily horse crossing. Of course, the constancy of such a unit of measure more than doubtful, although the batyrs of Genghis Khan, apparently, were quite satisfied with it. But Mongols even in head not came measure circle Earth! Will yours but With ancient astronomy, something is not so simple if, for example, an ancient Roman architect Vitruvius (I century BC) knew the periods of heliocentric (that is, around the Sun) revolutions planets better Copernicus.

An indirect argument in favor of the validity of our reasoning can be absolutely caveman level cosmological representations in early medieval Byzantium. Enlightened Byzantine Cosmas Indicopleutus (Kozma Indicopol), recognized specialist on medieval cosmography, thought what Universe represents yourself rectangular box, washed by waters great rivers Ocean. The vault of heaven is supported by four sheer walls. Stars, according to Cosmas, there is nothing else than the little carnations with which the lid of this box is stuffed, but four wind-producing angels are placed in the corners of this unintelligible structure. By the way, the said Cosmas lived in 6th century already a new era, that is, after 900 years after Aristarchus and 700 after Eratosthenes. But Byzantium is Eastern Roman empire that was once part of the enlightened Pax Romana, which, in turn, inherited Greeks. AT difference from Western Roman empire Byzantium not subjected devastating raids of barbarian tribes, and indeed the time since the fall of Rome (476 year) a little bit has passed - about 100 years. Okay, considering unconventional historical versions is not included in our tasks. These are just remarks, as they say, according to about...

So, more than 100 years before the beginning of the Christian era, astronomers managed to measure distance to the moon, and very accurately. What about other celestial bodies? How far are they from Earth? The already mentioned Aristarchus of Samos (IV-III centuries. before n. e.) tried calculate distance from Earth before sun, but suffered fiasco. The mathematical reasoning of the Greek astronomer was quite flawless, but the tools at his disposal were no good, so the magnitude turned out less true distances nearly in fifteen once. (However, many historians doubt the real existence of Aristarchus and, not without reason, believe that that the achievements of European astronomers of the 16th century are attributed to him.) The result of Archimedes was much better (2/5 of the actual value), but this is very alarming, since even Johannes Kepler in the 17th century could not cope with this task, the calculated by him the distance was even shorter. Be that as it may, the sky has moved into an utter distance, a Universe turned out much more, how could think the most daring minds antiquity.

After Hipparchus and Ptolemy, stagnation set in in the astronomical sciences. Stagnation continued over one and a half thousand years, up to before start XVI century, when Polish Priest Nicholas Copernicus proposed new model universe With motionless sun in center, received title heliocentric. According to this models, the planets revolved around the sun in regular circles, and their number decreased to six (Mercury, Venus, Earth, Mars, Jupiter, Saturn). The moon, strictly speaking, has lost the status of a full-fledged planet and turned into a natural satellite of the Earth. Although the model Copernicus was much simpler than Ptolemaic and gave somewhat better results, her on the throughout nearly 100 years Seriously not perceived. fracture happened in XVII century, when the Italian astronomer Galileo Galilei first managed to see through a telescope (which he himself invented in 1608) the satellites of Jupiter, followed by the great Johannes Kepler introduced amendments in scheme Copernicus. Having analyzed shiny observations Mars performed by his teacher, the Danish astronomer Tycho Brahe, Kepler concluded that the only geometric figure, which perfect answers this observations - ellipse. So, in the modified model of Copernicus, the planets began to revolve around sun on elliptical orbits a Sun moved in one from tricks this ellipse.

Moreover, Kepler found that between the average distances of the planets from the Sun and there is a simple mathematical relation between their periods of circulation. In this way, became possible calculate relative distance between sun and any from planets. Unfortunately, this did little, because the scheme proposed by Kepler (quite reliable and remarkably consistent with observations), there was no scale at all. One could say that, say, Saturn is located 10 times farther from the Sun than the Earth, but what this distance in kilometers is - a mystery shrouded in darkness. But if it were possible some way to calculate the distance between the Earth and any of the planets, astronomers the required scale would immediately appear in the hands. It was a matter of small - to come up with such way.

For definitions distances between heavenly bodies use phenomenon parallax. Parallax is a very simple thing. If you consider your own finger against a colorful background wallpaper right and left eye alternately, easily make sure that in that the moment you close one eye and open the other, the finger moves a little background distance. The closer the finger is to the eyes, the larger it will be. bias. The essence of the phenomenon lies on the surface: since the eyes are separated by some distance friend from friend, you see on the subject each eye under certain angle.

The same approach can be easily applied to celestial bodies. Of course, successively blink eyes, looking, let's say on the the moon absolutely meaningless because the she is located too much long away. BUT here if two astronomer, separated distance in several hundred kilometers, our natural satellite will be simultaneously observed at background of the starry sky, lunar parallax is easily detected. We just need to agree regarding which star observations will be made, and then the first astronomer will see the edge lunar disk on the one corner distance from in advance selected stars, a second, respectively, - on the otherwise. Farther - already a business techniques: if known bias Moon relatively stellar background and distance between observatories, then With help simple trigonometric functions can calculate distance before Moon.

AT progress such observations It was established, what magnitude lunar parallax is 57 minutes of arc, or about 1 degree of arc (a full circle is 360 degrees; There are 60 minutes in one degree and 60 seconds in a minute). Offset at 57 minutes of arc is very easy to measure, since it is approximately equal to two apparent diameters complete Moon. Distance, computed With help parallax, showed good coincidence with the numbers obtained by the old proven method - by the earth's shadow in time lunar eclipse.

But there was a problem with the planets. The problem is they are too far away. therefore, the parallactic shift is so small that it could not be measured up to the beginning of the 17th century. The problem was successfully solved after the invention of the telescope in 1608 year. In second half XVII century two French astronomer, jean Richet and Giovanni Cassini (Italian by origin), calculated by the parallax method distance from Earth before Mars. Observations were carried out simultaneously in paris and French Guiana. Kepler's model finally received the desired scale, after which all other distances within the solar system could be calculated without difficulty. AT in particular Cassini determined what distance from Earth before sun is 140 million kilometers. For XVII century this is very not bad accuracy, So how he wrong Total for 10 million kilometers. Technology did not stand still, and in the first half of the XVIII century Cassini's result was corrected to 152 million kilometers (the current value is 149.6 million kilometers). This value subsequently called *astronomical unit* (a. e.) and become wide apply in quality his kind interplanetary miles.

Sunny system acquired impressive dimensions: for example, distance from The sun to Saturn is almost one and a half billion kilometers, almost ten times more than to Earth. And when the English astronomer William Herschel discovered in 1781 Uranus (this planet is not visible to the naked eye, so the ancients did not know anything about it existence), Sunny system straightaway same grew up nearly twice (between Uranus and The sun lies about 3 billion kilometers). In 1846 the French astronomer Urban Joseph Le Verrier discovered Neptune, and the American Clyde Tombaugh in 1930 discovered Pluto, the ninth and last planet. Thus, the solar system again doubled in size, for Pluto is separated from the Sun by almost 6 billion kilometers, or about 40 astronomical units. And its diameter will, respectively, be equal to 12 billion kilometers (80 AU). A beam of light that flies 300,000 kilometers per second and travels in a second with quarter before Moon and per eight minutes before sun, would need near 12 hours, to cross her from end in the end.

Let's try more visually introduce yourself relative scale solar systems. If a portray Sun in billiard room ball (about 7 centimeters in diameter), then to Mercury - the planet closest to the Sun - will be on such a scale almost three meters (280 centimeters), and to the Earth - a little more than seven and a half meters. The giant planet Jupiter will move to a distance of about 40 meters, and Pluto will have to commit decent walk because the he will be lie in 300 meters from Sun. The dimensions of the Earth on this scale will be only 0.5 millimeters, so to see such a speck of dust can only be a person with good eyesight. So it's better to make it a little more: let magnitude Earth will be correspond size standard wrist hours. Then on this scale the diameter of the Sun will be equal to twice the average human growth, and the distance between the Earth and the Sun will be 400 meters. Pluto will be and at all not see because he retire on the distance in fifteen kilometers.

However, the orbit of Pluto is by no means the most distant point in the solar system. When in 1684 year great English scientist Isaac newton opened mine famous law universal gravitation, according to which bodies are attracted to each other with force, directly proportional to the product of their masses and inversely proportional to the square of the distance between them, Kepler's model acquired a mathematical justification. Scientists have received arms reliable tool, allowing calculate any orbit, even if body observed on a small segment of its trajectory. Astronomers have long been occupied with comets - caudate guests, time from time emerging on the firmament. Friend and contemporary Newton, Edmund Halley saw a distinct periodicity in the behavior of some comets. and suggested what they are moving around sun on very strongly elongated orbits (ellipses with a large eccentricity, as astronomers say). Halley calculated the orbit one of these comets and predicted that it would return again in 1758. 16 years after his of death prediction halley came true: comet really appeared on the sky in specified them year and With since then wears his name, regularly returning every 75 or 76 years.

At its perihelion point (closest to the Sun), Halley's comet is inside orbit of Venus, and at aphelion (the point of maximum distance from the Sun) goes far beyond orbit Neptune - on the 5 With superfluous billion kilometers. However exist So called long-period comets, which apply on so elongated orbits what are returning to sun times in several centuries a then and millennia. AT In the middle of the last century, the Dutch astronomer Jan Hendrik Oort suggested that that far beyond the orbit of Pluto lies a huge cloud of comets, from where they come from time to time penetrate in neighborhood Sun. AT such case diameter solar systems maybe achieve 1000 billion kilometers and even more, or dozens thousand astronomical units. Today, the Oort hypothesis has practically become a theory. Detailed story about planets solar systems and heavenly bodies lying per orbit Pluto you, reader, you can find in chapters "Ring around Sun" and "Nine or ten?".

So to early XVIII century size question solar family was practically resolved (of course, without the last three planets that were discovered later). Left deal with the fixed stars, finding out once and for all what they are. What they like this: Total only points on the spherical firmament, lying at most borders solar systems, how believed ancient, or huge heavenly body, remote on the monstrous distance? The parallax method, which has proven itself remarkably well when calculating the distances between the planets, obviously did not work here, since none of them stars not managed register any conspicuous offset. Even if observers were separated by a distance equal to the diameter of the Earth, the gap between adjacent stars not changed neither on the iota.

However, there was one more possibility. The diameter of our planet does not reach and 13 thousand kilometers, but after all, the Earth, as you know, does not rest in place, but rapidly flies through the void around the sun. Opposite points of the earth's orbit are separated by space of almost 300 million kilometers. The solution suggested itself: if one evening to put the position of the stars on the map, and then do the same exactly after half a year, then the astronomer will observe the starry sky from two points separated by a huge distance, superior in 23 thousands once complete length earthly diameter. Relevant way must increase and parallax. Per year star describe tiny ellipse - his kind image terrestrial orbits in miniature, a angular distance from the edges this ellipse before his center how times and will be parallax stars.

For planets similar method not good because they winding whimsically across the sky on the throughout of the year, masking topics most parallax bias, called movement Earth. Separate own traffic planets from her parallax - a task overwhelming complexity. But the stars are practically stationary throughout the year, so discover they have a parallax shift is quite real. The logic seems to be flawless, but no stellar parallaxes could be detected. It has been in the yard for a long time XIX century, but astronomers, no matter how they fought, could not determine at least a sensitive bias neither at one star.

The situation was becoming very unpleasant. Of course, one can always assume that all stars without exceptions are on the one and volume same distance from Earth. Then, of course stellar parallax not will be, because the parallax bias arises only in volume case, if we compare position close subject With position relatively distant. However hypothesis solid firmament, or thin spherical shell, on the surface of which the stars are located, looked very doubtful. The stars vary quite a lot in brightness, and to make sure of this, enough simply look on the nocturnal sky. classify them on this parameter the ancient Greeks learned, dividing the entire stellar population into 6 magnitudes (1st star 100 times brighter than a 6th magnitude star). It is clear that with the invention of the telescope star regiment arrived, as it became possible to observe stars that are indistinguishable naked eye. The number of stellar magnitudes immediately grew considerably. It was reasonable assume that the true luminosity all stars lies in pretty narrow limits, and the difference in their apparent brightness is due solely to distance. On the other hand, it is forbidden It was reset co accounts and opposite consideration: all stars lie at about the same distance from the Earth, but they shine in completely different ways, like light bulbs greater and less power.

Concept equidistance stars With crackling failed when astronomers guessed it apply to antique stellar directories. First systematically Hipparchus began to catalog the stars, and Ptolemy continued his work, leaving to posterity fundamental treatise "Almagest", in which fixed coordinates 1000 With superfluous stars. AT 1718 year already familiar us Edmund halley, studying stellar sky, unexpectedly discovered that at least three stars (Arcturus, Procyon and Sirius) are not at all where they were noted by the ancient Greeks. The discrepancy was so great that mistake not could to be and speeches: for example, Arcturus defended on the whole degree from specified in "Almagest" points. We remember that a degree is a distance twice the diameter. full moon. It remained to assume that the stars, like the planets, have their own movement, only them traffic incomparable slower if Arcturus took more one and a half thousand years, to shift to one degree.

The search for stellar parallaxes continued, but the first success came to astronomers only in 30s years XIX century, when telescopes and astronomical tools become much more perfect. AT 1838 year German astronomer Friedrich Wilhelm Bessel managed define parallax 61 swan, year later published their results Englishman Thomas Henderson (he studied position of Alpha Centauri) a 1840 Russian astronomer Vasily Yakovlevich reported on his observations of the bright star Begi Struve. Justice for the sake of should would give away palm tree championship exactly Struve, because he finished the work before everyone else - in 1837, but he was somewhat late with publication. The stellar distances turned out to be unimaginably huge. Even the closest Sun star - Alpha Centauri (in fact, it is a triple star, and closest to the Sun lies third, weak her component - Proxima, what translated how "nearest") located on the distance 4.3 light of the year. interplanetary verst - astronomical unit is no longer suitable for such open spaces, so astronomers use the interstellar a mile is a light year. *Light year* - is the distance that a beam of light traveling from speed of 300 thousand kilometers per second, overcomes in a year. Remember that light the beam takes only 8 minutes to reach the Sun, and about 6 hours to rush before Pluto a before nearest stars he has to crawl over four years. If a whatever, you can try to express is the distance in kilometers: since one light year approximately equal to 9.5 trillion kilometers, then the distance to Proxima Centauri is near 40 trillion kilometers (40,000,000,000,000 km).

If we recall our model with a billiard ball in place of the Sun, the Earth in seven half meters from it and Pluto at a distance of about 300 meters, then on this scale distance between sun and nearest to him star will be dress nearly 2000 kilometers. BUT in models, where Earth was magnitude With wrist watch, a Pluto was in
fifteen kilometers from her get there before proxima centauri will be very problematic because the this is distance will be near 100 thousand kilometers - two With half around the world travels. More more visual example invented one Moscow lecturer. He took a piece of chalk and declared it "planet Earth", and a board hanging on the wall - Sun. From the blackboard to the chalk was only one meter, designed to depict the astronomical unit - 150 million kilometers, separating sun and Earth. "How in this scale to the nearest star? the lecturer asked the audience. The audience became timid speak out. Someone suggested that the star would be in a nearby lane, but most resolute stood for the outskirts of the city. Meanwhile, the star was in Yaroslavl (or any friend city, remote on the 300 kilometers). More once we emphasize what this is nearest to the sun star.

Besselevskaya 61 swan turned out more farther - in 11.1 light of the year, a before run which was studied by V. Ya. Struve, was 27 light years. This is the scale of stellar distances. After definitions first parallax at nearest stars received wide Spread more one interstellar mile - *parallax second,* or parsec. *Parsec* (pc) - this is distance, on the which star at her observation With opposite points Earth's orbit changes its apparent position by one arc second of arc. Or more simpler: the distance from which the earth's orbit is visible at an angle of one second of arc. One parsec equals 3.26 light of the year, 206 265 astronomical units or 30.857 X 10^{12} kilometers (slightly more thirty trillion kilometers). Distance before proxima centauri is 1.3 parsec, before 61 swan - 3.4 parsec, a before run - 7.8 parsec. suggested conclusion, what stars - by no means not dimensionless points on the firmament, a gigantic sun, in everyone similar our native luminary, only remote monstrously long away, on the distance measured by many light for years.

By calculating the true distance to the star, you can calculate its luminosity, that is, not visible stellar value, a genuine strength her Sveta, which received call absolute stellar size. Quite possible and reverse procedure: mentally by placing a star at any arbitrary distance, one can determine how bright she is will be seem earthly observer. Absolute stellar magnitude called brightness stars on the distance in ten parsec (32.6 light of the year); of course stars distributed unevenly in space, but if we line them up on a specified distance, then we can compare them valid luminosity. Our Sun on the a distance of 10 parsecs would be a very faint star with an absolute magnitude of 4.9, and Sirius is the brightest star in our sky - it would shine almost the same as it shines on his place (2.7 parsecs, or about 9 light years). Its absolute magnitude is 1.4, from what follows, what true luminosity Sirius exceeds sunny in 25 once. Of course, this is far from the limit: the blue giant Deneb (we will talk about the classes of stars in next chapter) exceeds on luminosity Sun in 270 thousand once; he not looks especially bright only because it is very far from us (more than 3 thousand light years).

In other words, the apparent brilliance of a star says nothing about the amount of light which she radiates. The sun shines extremely brightly, because it is located literally in two steps. Sirius is about four times brighter than Vega from the constellation Lyra, and the guide The North Star is the dimmest of them (six times fainter than Vega). However, if we produced reassessment of values and lined up these stars on the the same distance from Earth, then the Polar Star would confidently take the first place, and the second place would be Vega, on the third - Sirius, but magnificent Sun became would hopeless outsider.

When in the middle of the century before last it was possible to determine the distance to the nearest stars, the question immediately arose of how far they extended. naked eye can see near six thousand stars, but when Galileo looked on the sky in my a primitive spotting scope, he immediately discovered that the stars were poked far more densely. It's just that many members of this glorious family are so weak that you can't see them. without the aid of a telescope there is no possibility. Modern astronomical technology allows you to distinguish stars of the 25th magnitude. In addition, already in the time of Herschel it became clear that the stars are distributed in space very unevenly. If you look at the sky dark moonless night, you can see a faint foggy glow encircling the entire firmament from horizon to horizon. To unfortunately bright urban the lights not allow make out his how should (electrification, With points vision astronomer, generally a dubious blessing), but somewhere in the wilderness you can easily see soft luminous dairy strip, crossing nocturnal sky. ancient Greeks called her galaktikos ("milky, milky"), and the Romans - via lactea, which literally translates means "milky path". Origin this titles associated with antique the myth of jet milk, which splashed on the sky from chest goddesses Hera, wives Zeus when she is pushed back baby on my own Hercules.

There are far more stars in the direction of the Milky Way than in any another parts firmament, that's why Herschel reasonable suggested what stars not distributed evenly, a collected in compact structure, having form biconvex lens. According to Herschel, our star system (later it became call the Galaxy) could contain about 300 million stars and be 15 thousand light years (let's not forget that the first stellar parallaxes were measured only through 16 years after of death Herschel). Today we we know what our galaxy *milky Path* (or just *Galaxy* with a capital letter) is much larger: its diameter is 100 thousand light years, and the number of stars reaches 200 billion (however, the number stellar population, according to estimates by various authors, varies widely - from 150 to 400 billion stars).

Here necessary do small retreat and tell to the reader what these parameters were calculated in this way. Since the parallax shift with large labor succeed measure even near the nearest stars, parallax detection at the objects more than 100 light-years away, becomes an almost impossible task. Parallax is a value derived from the proper motion of a star, so it is clear that that the farther away a star is, the more difficult it is to catch its movement across the sky. Not going into in details, let's say what astronomers helped out So called Cepheid scale. Cepheids are called pulsating variable stars, which are strictly periodic change their brightness by one or two magnitudes (radiation power increases by 2.5–6 once on comparison With minimum). Actually various variables stars exists lots of; one of the most famous is the red giant Omicron Ceti, discovered back in late 16th century by the German astronomer David Fabricius. This star is several times changes its brilliance with a period of about 11 months, so she was called Mira (translated from Latin - "amazing"). However greatest meaning for astrophysicists have short-period variable stars with a period of from a day to a month (usually about weeks). This is exactly the Cepheus delta, changing brightness with a period of 5.37 days, which gave his name for everything family similar stars.

AT early of the past century American astronomer Henrietta Leavitt discovered correct relationship between luminosity and period of some Cepheids. The more there was a period, the more energy the star radiated per unit time. Having calculated the power radiation according to the "period - luminosity" dependence, scientists were able to calculate the distance to Cepheids. First, relative distances were established (how many times one star closer or farther than another), and then absolute ones, taking into account the radial velocity of Cepheids (in spectrum of a star approaching or receding along the line of sight, a shift occurs spectral lines). Astrophysicists got reliable scale. BUT at all recently on the astronomers were helped by supernovae of a certain type (type 1a), the luminosity of which lies within very narrow limits. About these stars, called "standard candles", detail told in chapter "And darkness came."

By the beginning of the 20th century, the world had expanded unimaginably. It became finally clear that The sun is one of many hundreds of billions of stars that inhabit our Galaxy, and far from the most remarkable. In star nomenclature, it is listed as an ordinary yellow class G dwarf. Yes, and lies, moreover, by no means in the center, as he believed, for example, Herschel, and on the periphery of the Milky Way, in one of its spiral arms - 26 thousand light years from center galaxies (about eight kiloparsec). clearly imagine these overwhelming expanse is not easy. If we shrink the entire solar system to the size of a grain of sand, then the nearest star Proxima Centauri will be on this scale at distance of one meter, and the distance to the center of the Galaxy will be almost 9 kilometers. If we recall the model with a billiard ball in place of the Sun, the dimensions of the Milky Way will equal 60 million kilometers - a value quite comparable with the distance from Earth to the Sun.

However, the universe is not limited to the Milky Way galaxy. If we could leave her limits, before us swung open would immense empty space, impenetrable coal blackness, devoid of any noticeable objects. And only on about 200 thousand light-years from our star island, we would find two ragged foggy education wrong forms - big and Small Magellanic clouds. They are Good visible on the sky Southern hemisphere in form two whitish spots and look like isolated fragments of the Milky Way. For the first time described one of the participants around the world swimming fernana Magellan. direct relations they do not have to the Milky Way: these are two independent small galaxies, quite poor stars. The Small Magellanic Cloud lies 160,000 light-years away, and The big one is pushed even further - by almost 200,000 light-years. Although the Magellanic clouds are noticeably smaller than the Milky Way in size, very curious objects. For example, the star S Doradus is located in the Large Magellanic Cloud, possessing greatest famous luminosity. unarmed eye she is not visible because what It has 8th stellar value, but her absolute luminosity surpasses sunshine 600 thousand times! And in the Small Magellanic Cloud there are hundreds of already acquaintances us cepheid, which systematically studied Henrietta Leavitt in early of the past century.

If a would we looked With such distances on the our own galaxy, then would see an impressive spiral disk, vaguely resembling a furiously spinning whirlpool (shape biconvex lenses or spindles she is acquires at look With ribs). However milky Path and Magellanic clouds - this is more long away not all. AT 2 With half million light years from milky Ways lies spiral galaxy andromeda, much superior our on mass and quantity stars. She is visible to the naked eye as a faint asterisk of the 5th magnitude and is listed in the Messier catalog under number 31, so it was called M31. (Charles Messier - the famous French astronomer, one from first started make up catalog nebulae and stellar clusters.)

Andromeda Galaxy, Milky Way, Magellanic Clouds, Spiral in Triangulum (MZZ) and lots of galaxies slightly less (general number near 40) are included in compound So called *the Local Group* with a diameter of over 3 million light years. Within 10 Mpc (megaparsec, that is, millions of parsecs), or more than 30 million light years, scattered about a dozen similar groups. And at 15 Mpc (almost 50 million light years) lies a large cluster in the constellation Virgo, numbering several thousand galaxies. So the way our local Group belongs to more more large-scale structure, commonly referred to as a local supercluster of galaxies. Its diameter is 30 Mpc, a thickness - near ten Mpc (100 and thirty With superfluous million light years respectively). Center this gigantic galactic clouds is the aforementioned cluster in Virgo.

The Milky Way galaxy huddles at the very edge of a local supercluster. And also farther, on the distance in 90 Mpc (check goes already on the hundreds million light years), located much more large accumulation in constellation Hair Veronica, in compound whom included over 10 thousand galaxies. By all appearance, it represents yourself part of another giant galactic supercluster, which recently Dozens are open. Thus, they crown the hierarchy of structures of our *Metagalaxies* (of the observable part of the universe). Only at distances of the order of many hundreds million light years universe can consider how relatively homogeneous structure, which contains dozens billion galaxies. Modern astrophysics has high-precision perfect equipment that allows you to conduct observations in the widest range of waves - from meter radio waves to gamma rays. Apart from traditional optical telescopes wide apply infrared and radio telescopes, as well as X-ray and gamma-ray detectors. Rapidly developing neutrino astronomy. Scientists have access to unimaginable distances measured 10-12 billion light years, when the world was still young and fresh, and the first galaxies barely managed form. So the way dimensions observable parts Universe can estimate approximately at 6 thousand megaparsec.

When we look at distant stars or galaxies, we should keep in mind that we moving backwards along the time axis. If Sirius is about 9 light years away, we see its the way it was 9 light years ago because light has a finite speed distribution. Rays red giant Betelgeuse from constellations Orion set off in way back in the Time of Troubles, when Boris Godunov sat on the Russian throne. Ball star clusters at the center of the galaxy will take us back to the last ice age, and the light the Andromeda Nebula was emitted at a time when our ape-like ancestors stood on two legs and turned the first stones. The most distant objects in our universe send light from era, remote in past on the many billions years. solar systems and planets Earth then more not was in remember.

In order to personally, in living images, estimate the size of the observed part of the Universe, or metagalaxies, mentally reduce earthly orbit (her diameter 300 million kilometers) to the size of the inner electron shell in the classical model of the atom Bora (her radius equals 0.53×10^{-8} cm). Then nearest star will accommodate although and on the small, but quite macroscopic distance of 0.014 millimeters, the distance to the center of the Galaxy will be 10 centimeters, and the diameter of the Milky Way will be 35 centimeters. The Andromeda Galaxy will recede by as much as six meters from the Bohr atom, and distance to the central part of the cluster of galaxies in the constellation Virgo, which includes our The local group will be about 120 meters. Radio galaxy Cygnus A (before it 600 million light years) "run away" to this scale on the two With half kilometers a before distant radio galaxy 3C 295 will have to walk and walk - after all, 25 kilometers. All in all, terrestrial ball huge how With pathos one teacher said elementary grades...

Star freak show

- Yes... We live we live - a why? Secret centuries. And unless comprehended anyone thin filiform the essence of the luminaries?

Victor Pelevin

outside any doubts, the most noteworthy and common objects our The universe is the stars, so it makes sense to start talking about its "inhabitants" with them. World stars strikes their variety. Among them there is giant stars and dwarf stars, collectivist stars, preferring go astray in flocks, and star anchorites living in splendid isolation. Many stars form so-called multiple systems of two or three stars that revolve around a common center of gravity on the relatively small distance friend from friend. Alone stars similar dark ghosts, because they shine in the infrared range, while others shine in tens and hundreds thousands of times brighter than our sun. And only in one parameter - by mass - they are not very vary greatly among themselves: from 1/10 of the mass of the Sun to 100 solar masses. Stars almost like people they are born, grow up, grow old and are dying. But if alone go to other world quietly and imperceptibly, then the death of others is accompanied by grandiose cosmic cataclysms, received title explosions supernovae. Such stars visible on the distances of many millions of light years, and their brightness exceeds the richest imagination: intolerable shine supernova dwarfs cumulative shine hundreds billion stars of the entire galaxy.

How known nothing not forever and to stars this is applies in complete measure. Each timed out. Some stars live bright and festively, burning down in a matter of millions years. When dinosaurs roamed the Earth, they didn't exist yet. The ephemeral existence of these mayflies fit in one a short galactic instant. Other lead measured unhurried Existence and will live for a long time: time life stars, a little less massive how Sun, maybe achieve 25 billion years (our Universe was born only about 14 billion years ago). The sun lit up about 5 billion years ago and today is "a man in the prime of life," as Carlson used to say. Like lyrical hero Dante it managed pass the earthly life Total only before half. Some stars destined not easy fate: when they burn down to the ground his nuclear fuel, then turn into in black holes - amazing objects, possessing very strange and even frightening properties. Path to center black holes - this is descent in hell, road without return, because the strength gravity on the her surfaces reach such magnitudes that even the light is not able to get out. Monstrous gravity like severe tombstone stove forever and ever fences off black hole from our peace. However, about blacks holes we in his time yet let's talk.

The first thing that catches your eye, even with a cursory glance at the night sky, is a distinct difference between stars in brightness and color. The ancient Greeks, as we remember, smashed the entire stellar audience into six classes, which are called stellar magnitudes. Stars stars of the first magnitude are 2.512 times brighter than stars of the second magnitude, and so on. In this way, stars sixth quantities weaker stars first quantities in 100 once. Apart from visible stellar magnitudes, there are absolute magnitudes, which I already wrote about in the previous chapter, so I won't repeat it. In fact, the absolute magnitude is the same the same as the luminosity of a star (it is usually expressed in units of the luminosity of the Sun and denoted by the letter L), that is, the total amount of energy emitted by a star per unit time. The stars vary greatly in this parameter. Let me remind you that the luminosity of Deneb exceeds the solar one by 270 thousand times, and the brightness of S Dorado in the Large Magellanic cloud exceeds the luminosity of the Sun by 600 thousand times. Among other bright stars of our sky can be mentioned Antares (alpha Scorpio), Betelgeuse (alpha Orion) and Rigel (beta Orion), luminosity which exceed sunny in four thousand, eight thousand and 45 thousand times respectively. On the other hand, the luminosity of dwarf stars can, in turn, yield solar luminosity in thousands and tens thousand once.

Only very bright stars can see the difference in color with the naked eye. Let's say Antares and Betelgeuse will red, Chapel - yellow, Sirius - white, a Vega
- bluish-white. But a small amateur telescope or even a decent field binoculars will significantly improve the quality of the picture. The color of a star, and hence its spectrum determined by the temperature of its surface layers. At a temperature of 3-4 thousand degrees Kelvin star will be red, at 6–7 thousands degrees acquires distinct yellowish hue, and hot stars with a temperature of 10-12 thousand degrees shine white or bluish light. AT contemporary astronomy there are reliable and quite objective methods for measuring the color of stars, with the help of which the magnitude under name "index colors". To each meaning indicator colors corresponds definite spectrum type.

Received allocate seven major spectral classes which designate Latin letters O, B, A, F, G, K and M. For greater accuracy, each spectral class divided into 10 subclasses (from 0 to 9 with increasing downward temperature). So Thus, a star with spectrum B9 will be closer in spectral characteristics to spectrum A2 than, for example, spectrum B1. Stars of classes O - B are blue (surface temperature - approximately 100 - 80 thousand degrees), A - F - white (11 - 7.5 thousand degrees), G - yellow (about 6 thousand degrees), K - orange (about 5 thousand degrees), M - red (2-3 thousands degrees).

Our Sun belongs to the spectral class G2 (the temperature of its surface layers - about 6 thousand degrees) and is considered, no matter how insulting, a dwarf yellow star. However, the size of this dwarf is quite decent - the diameter of the Sun is about 1.4 million kilometers.

Some stars may periodically change mine shine. AT first chapter told about cepheids, pulsating variables stars, which sometimes called
"beacons of the Universe", because thanks to them it was possible to build a reliable scale, with the help of which astronomers have learned to determine the distances to distant stars and other galaxies. Cepheids are yellow supergiants with a surface temperature of about the same as the Sun. But they shine much brighter, because the power of their radiation surpasses sunny in dozens thousand once. periodic change shine stars

of this type is associated with complex physicochemical processes in their depths, therefore they are usually called true, or physical, variables. Star of the World from the constellation Kita is also among the real variables, although the period of change in brightness in her much more and is about eleven months (at cepheid - from days before months).

However, there are variable stars whose brightness fluctuations are in no way related to features them internal buildings. An example such star is Algol (beta Perseus), which in antiquity called "eye devil" and "ghoul". Her brightness changes by a whole magnitude every three days without three hours. The Greeks placed beta Perseus into the head of Medusa Gorgon - a terrible fanged monster in a female form and with snakes instead of hair. The gaze of this winged creature turned all living things into stone. Algol applies to number So called eclipsing double stars, because what the reasons the variability of its brightness is fundamentally different than that of the delta Cepheus or the omicron Cetus. Around Algol draws weak star - second component double systems, orbit which lies in one plane With terrestrial orbit. When she is turns out between Algol em and the Earth on the line of sight of an earthly observer, then partially overshadows it. In this way, intensity radiation Algol in reality not intensifies and not is weakening a remains strictly constant. Quite simply on the way dissemination light rays periodically an obstacle arises.

It is reasonable to assume that since the surface temperature of red stars of the spectral class M is more than two times smaller than the sun, then they should shine very weakly. However, in reality, everything turned out to be far from being so elementary. Some class stars M (let's say "flying" Barnard) really smolder barely, although they are at all close to the Sun (the distance to Barnard is about 6 light years). But many others, certainly falling in the same spectral class, burn very brightly, despite on the significant remoteness from Sun. For example, Antares in scorpio and Betelgeuse from the constellation Orion - classic red stars - not only have a visible less than unity, but also have a large intrinsic luminosity. Power Radiation from Betelgeuse is 8,000 times greater than the sun's. It is clear that such a high luminosity relatively cold stars maybe to explain only her gigantic sizes. And although the surface of the red giant is heated to only 2-3 thousand degrees, total intensity light flow will be very significant on comparison With Sun. Let a square kilometer of Betelgeuse's surface shine relatively weakly, but there are orders of magnitude more such square kilometers on the body of a star, therefore power her radiation in many times exceeds solar.

In 1920, the diameter of Betelgeuse was measured. Although the stars, even in the most powerful telescopes are seen as dimensionless points, an ingenious method has been devised to calculate them sizes. A business in volume, what rays Sveta, coming to earthly observer from opposite points of the stellar disk (which we do not perceive as a disk) form, topics not less, some corner between yourself. Of course measure his value directly impossible, but light rays, overlapping friend on the friend, interfere with each other, so that with the help of a special device (interferometer) you can measure result similar additions and calculate value angle. Knowing this corner and distance before stars, maybe without special labor calculate her valid diameter. Of course, the method has its limitations (the angle should not be vanishingly small), but in many cases he properly works and very not bad yourself recommended.

Calculated so way diameter Betelgeuse struck imagination. It turned out that it is almost 350 times the diameter of the Sun and is approximately 500 million kilometers. Recall to the reader what orbit Mars lies in 220 millions kilometers from the sun. If it were possible to place this star in the place of our luminary, the surface layers of Betelgeuse's photosphere would extend far beyond the orbit of Mars, and all four terrestrial planets (Mercury, Venus, Earth and Mars) would sink into stellar bosom. Surface Betelgeuse will be nearly in 120 thousand once more surfaces sun, that's why hardly whether costs be surprised, what her luminosity in several thousand once surpasses the sun. The volume of this red star is 40 million times that of Sun. Despite such a fantastic size, the mass of Betelgeuse is estimated at only only 12–17 solar masses, that is, its average density should be negligible. Red supergiants, inside which may fit several planetary orbits solar systems, can compare With huge bubbles. If a average density sunny substances is equal to about 1.4 g/cm3(almost in one and a half times more density water), then in such monstrously swollen bubbles it will be millions of times less than in air.

Betelgeuse is by no means unique among the stars. There are red supergiants so unimaginably huge, what stars like Antares or Betelgeuse seem beside Withthem mere crumbs. For example, Epsilon Aurigae is larger than Alpha Orion.at least five times, but we do not even see it, because the radiation of this monster nearly entirely lies in infrared areas spectrum. discover his managed due to presence bright satellite, which the periodically eclipsed invisible star. Epsilon Aurigae is an infrared supergiant with a diameter of 3.7 billion kilometers. If you place it in the place of the Sun, it will easily "swallow" the first 6 planets (Mercury, Venus, Earth, Mars, Jupiter and Saturn) and will fill the solar system up to to the orbit of Uranus. Another star of this type - VV Cephei A - is only slightly inferior in the size of his companion from the constellation Auriga. Its diameter is greater than the diameter of Betelgeuse more than three times. The search for invisible stars is associated with great difficulties, since the earth's atmosphere is almost opaque to infrared rays; in addition, own thermal radiation Earth extinguishes warm, coming from space. Tem not less managed measure temperature some stars, which shine in infrared range. She is located in within 800 - 1200 degrees Kelvin what, certainly same, very few: 800 degrees - this is only just temperature red heat. Dark and cold supergiants like VV Cepheus or epsilon Charioteer must to be empty sparse worlds, because their stuffing is smeared over a colossal volume. If by some miracle managed to transfer the substance of these stars to the earth's laboratory, its average density almost not would be different from vacuum.

Kohl soon in nature there are red giants and supergiants, naturally suggest that there must be red dwarfs that fall into the same spectral class M. Let us recall at least Barnard's "flying" star, rapidly moving across the sky at a speed of more than 10 arc seconds per year. This is a lot because the proper motion of stars is measured, as a rule, by much smaller values (about one second per year or less). An outstanding athlete owes its name to American astronomer Edward Barnard which the opened her in 1916 year. Red dwarfs, noticeably inferior in mass to the Sun, are by no means bubbles, but quite weighty complete stars. Moreover, very often they are much denser than our star. For example, red dwarf Kruger 60V easier sun Total in five once, although his volume is 1/125 of the sun. Therefore, its average density should be equal to 35 g/cm3, which is 25 times the density of the Sun (1.4 cm3) and one and a half times the density platinum. Even such solid heavenly body, how our native planet, It has middle density order 5.5 g/cm3(density stone breeds terrestrial bark is 2.6 g/cm3, a to the center of the Earth, it reaches a value of 11.5 g / cm3), that is, it is inferior to Kruger in six seconds superfluous once.

AT brackets note what density all heavenly tel (and extremely sparse gas bubbles like Antares and Betelgeuse here too not exception) swiftly growing on direction to center. To Sun could stable exist, not collapsing under the action of gravitational forces, the density of its central regions should achieve quantities order 100 g/cm3, what exceeds density platinum in five once. It is clear that in the center Kruger 60V similar indicator for extreme measure for two order more.

However, the density of red dwarfs is nothing compared to white dwarfs. White dwarfs - this is small and very hot stars, representing yourself final stage of evolution heavenly luminaries like our sun. Their temperature surface layers varies widely - from 5 thousand degrees for the "old" cold stars up to 50 thousand in "young" and hot. They are comparable in weight to the Sun, but their diameter, as a rule, does not exceed the diameter of the Earth (about 12,800 kilometers). Thus, their average density reaches values of the order of 106 g/cm3 and exceeds sunny in hundreds thousand once. One cubic centimeter substances white dwarf maybe to weight several tons. The first white dwarf was open in 1844 year Friedrich Bessel when he unexpectedly discovered anomalies in the motion of Sirius - most bright stars our sky. His trajectory on incomprehensible reason periodically deviated from the average position, so Bessel suggested that Sirius enters double system, then there is It has massive satellite star, a both luminaries apply around a common center of mass. In 1862, in the vicinity of Sirius, they managed to make out a dim speck, and since then the bright component of this binary system has been named Sirius A, and his minor dark neighbor got title Sirius V.

Sirius AT - long away not most small representative populations whites dwarfs. Since its luminosity is 300 times less than the sun, and the surface temperature reaches 8000 degrees Kelvin (temperature sun - 5800 degrees), does not amount to much labor calculate its dimensions. Sirius radius The must to be about 20 thousand kilometers (5 thousand kilometers less than Neptune, but three times more than the Earth), and since its mass is 95 % sun mass, then average density his substances equals 105g/cm3.

Of course, Sirius B is by no means an exceptional phenomenon. Was soon discovered superdense satellite of Procyon, almost twice as light as the Sun, and then the finds poured out as if from cornucopia. To date, quite a lot of white dwarfs have been discovered (although search these small dim stars conjugated With considerable difficulties), and on preliminary estimated on the them share account for several percent stars our Galaxies.

Despite the monstrous spread of the stellar population in terms of the density parameter - from almost complete vacuum to values comparable with the density of the atomic nucleus, the masses of stars differ not very much - from 0.1 solar masses to 100 solar masses. In this way, the heaviest star is only a thousand times more massive than the lightest. And you should have in mind that at the extreme poles of the scale there are relatively few stellar audiences, So how weight the vast majority of stars fluctuate within 0.2–5 solar wt. Weight - extremely important characteristic, because the defines not only stellar modus vivendi, but also its sad ending, and in a certain sense even posthumous destiny stars. But about evolution we are the stars in his time let's talk separately.

BUT how star weigh? If a co luminosity, indicator colors and spectral class that determines the chemical composition and temperature of the surface of a celestial body, we somehow figured out how to determine its mass? Indispensable and irreplaceable the instrument in such cases is the double stars already familiar to us. The fact, that it is almost impossible to measure the mass of a single star. Of course, the intensity brightness and spectrum can tell a lot, because they depend on the mass, but still I wanted to to know this value for sure. Fortunately, staunch anchorites like our Sun are relatively rare, since most stars prefer to live in a friendly team. More often Total this is paired double systems, less often - triple and even quadruple. It is not easy to create a structure of three or four stars, because such systems turn out to be dynamically unstable. To make them stable required comply row conditions. Third component must address around close binary system in a sufficiently wide orbit, never approaching a distance less eight - ten radii internal "twos". He myself, in my turn, maybe to be double system, and then these two pairs will perceive each other as point objects. AT in the first case, we have a triple star, and in the second, a quadruple. Due to the features There are no processes of star formation in systems of greater multiplicity in nature. Double stars revolve around a common center of gravity - the so-called barycenter, since each of them pulls the blanket over itself, "rocking" the neighbor with its gravitational field. Therefore, if the periods of revolution of stars and the distances from them to the barycenter are known, it is not will be big labor definitely calculate mass each stars.

Should to tell several words about flat diagram "spectrum - luminosity" (or "temperature - luminosity"), because astronomers widely enjoy. Because the for the first time, diagrams of this type began to be used by the Dane E. Hertzsprung and American G. N. Russell, they are usually called Hertzsprung-Russell diagrams. On the horizontal axis of this diagram, from left to right, the spectral types are laid out from O to M, that is, in order decrease in temperature. On the vertical axis from bottom to top are luminosities, or absolute stellar quantities, on measure them increase. Regardless friend from friend Hertzsprung and Russell found an empirical relationship between temperature and luminosity. How rule star topics brighter how she is hotter although, certainly, there are and exceptions (remember red supergiants). But in average this regularity works at all not bad. That's why how to the left lies spectral Class researched stars on the horizontal axes (Consequently, how more her temperature), topics above she is climbs on vertical scale absolute stellar quantities (luminosity).

So the way majority stars settled down on diagonals in form wide a band running from the upper left corner of the diagram, where hot and bright stars lay, to lower right corner, inhabited cold and dim red dwarfs. This wide diagonal tape called the main sequence.

Stars, lying on the main sequences are located not anyhow how, but obey certain rules. Straightaway same came to light relationship between temperature stars and her radius, because the it turned out, what star With certain surface temperature cannot be arbitrarily large, and hence its luminosity also fit into some fixed parameters. In addition, luminosity is related to the mass of the star. If we go along the main sequence from the spectral types O - B before To - M, then masses stars continuously decrease. Let's say at stars class O masses reach several tens of solar, while in class B stars they do not exceed 10 masses of the sun. Our Sun is known to have a spectral class of G2, so it will be nearly in middle main sequences a little nearer to her right bottom edge. Stars of later mass classes are noticeably less than the solar mass; for example, Red dwarfs of spectral class M are 10 times lighter than the Sun. The physical cause of all these patterns succeeded understand only after creation theories thermonuclear reactions.

However, far from the entire stellar population falls on the main sequence. Red giants and supergiants (they are traditionally called red, although among they also have yellow stars) form a separate branch, which grows in a wide strip from the middle of the main sequence and goes to the upper right corner of the diagram. We already these stars are well known With great luminosity and low temperature surfaces. Against the background of the bulk of the stellar population of giants, there are relatively few. And in the bottom on the left corner of the diagram are white dwarfs - hot stars with low luminosity, what He speaks about them very small sizes. running a little forward, let's say what white dwarfs present yourself regular final stage evolution some stars. Thermonuclear reactions in their bowels have not been going on for a long time, and they are slowly cooling down. So, suggests itself conclusion, what and red giants, and white dwarfs - this is his kind production waste, certain stage evolution stars, left home subsequence. BUT because the questions life and of death - alone from most burning, it has come time closer познакомиться With birth and evolution stars.

According to modern concepts, stars are born inside gas and dust clouds, which start shrink under action own gravity forces. interstellar Wednesday only on the the first sight seems nothing not filled empty space, but in reality it contains significant amounts of gas and dust, which are distributed very unevenly. Most of the gas and dust is concentrated in galactic spiral arms, and here the so-called associations young stars, what is additional argument in benefit them birth from gas and dust clouds. In addition to molecular hydrogen and atomic helium, such clouds contain small particles of cosmic dust composed of heavier elements. And although no one has yet been able to trace all the phases of star formation from beginning to end, in himself general form this process can be imagined next way.

After segregation and seals fragment clouds comes phase his fast compression. Density clot swiftly growing, a his transparency steadily falls, therefore, the accumulated heat cannot leave it, and the clot begins to heat up. Radius such protostars much surpasses radius sun, but she is continues shrink, because what pressure gas and temperature inside clouds not in able balance gravitational strength. When temperature in center protostars reaches several million degrees, thermonuclear fusion reactions flare up in its depths. The temperature and pressure continue to rise, and there comes a point when they begin to effectively resist forces gravitational compression. protostar becomes complete star and enough fast "sit down" on the home subsequence.

To "run through" most early phase his evolution, star required relatively a little time. Speed appearance on the light depends from weight baby. Heavy stars born much faster lungs. For example, at our sun, on some estimates, gone on the this is a business about thirty million years, a stars, triple surpassing it in mass, jump out like a cannon - in just 100 thousand years. But in red dwarfs, the mass of which is an order of magnitude less than that of the sun, childbirth stretches for hundreds million years, but but and live such stars much longer. Weight stars determines not only the circumstances of her birth and the first steps in this world, but also leaves an imperious imprint on her entire subsequent fate. But first, let's deal with processes leaking in stellar bowels, which provide newborn comfortable Existence.

Any star represents yourself self-adjusting nuclear reactor, providing prolonged and stable production energy. AT stellar bowels thermonuclear fusion reactions are gaining momentum, during which hydrogen is converted into helium, and that, in turn, gradually transforms into ever heavier elements. The main nuclear cycle of a star is the conversion of hydrogen into helium, because hydrogen in percentage in its composition the most. For example, our Sun, safely lived in the white world for about 5 billion years, contains a little more than 80% hydrogen. Rest twenty % fall on the helium and other, more heavy elements, but helium, of course incomparable more. Transformation hydrogen in helium in mostly carried out through So called proton-proton cycle, a because the he very slow, it ensures stable burning of the star for 10 billion years. AT jungle physical and chemical processes, ongoing in bowels stars, we not climb, a we only note that the lifetime of a star on the main sequence (that is, its period relatively quiet existence) depends primarily on its initial mass. Our sun and similar to him stars destined long and measured life (not less 5 billion years), and red dwarfs will live more longer.

Any star represents yourself red-hot plasma ball (helium and hydrogen plasmas, as astrophysicists put it), and thermonuclear reactions play a dual role: firstly, they maintain pressure at the required level and temperature, which oppose gravitational compression a Secondly, enrich star with heavy elements. The average chemical composition of the outer layers of a star looks like something like this: for 10 thousand hydrogen atoms there are 1 thousand helium atoms, 5 atoms oxygen, 2 nitrogen atoms, one carbon atom and 0.3 iron atoms. Relative content others elements more less. However accumulation heavy elements (a without them the emergence of terrestrial-type planets and, apparently, life is impossible) most actively going on in massive stars, which perceptibly heavier Sun. Helium in centers such stars starts turn into in elements carbon cycle (carbon, oxygen, nitrogen and etc.), and they, in turn, are transformed in even more heavy elements down to iron. Our Sun is known to be a relatively small star. (yellow dwarf spectral class G2), and calculations show what if would it originally on the 100 % was from hydrogen, to him it took would not less twenty billion years, to reach contemporary ratios hydrogen, helium and others elements. Meanwhile, the solar "age" has no more than 5 billion years. What way sun managed so fast get rich heavy elements, if his massesfor is this clearly not enough?

To answer this question, you need to look what happens to the stars main sequences. How we remember being on the main sequences star stable radiates on the throughout long time and her position on the diagram
"spectrum - luminosity" does not change. However, the hydrogen fuel consumption supporting thermonuclear fusion reactions in the depths, is not the same for different stars. Stars comparable to The sun by mass, they live very economically, so they have enough hydrogen reserves for a long time. Red dwarfs are even bigger misers: carefully counting every penny, they will live twice, and even three or four times longer than our Sun. But massive stars are great spenders and motes: the heaviest of them will be on the main sequence only several million years. Stormy life in young years leads to early old age.

What happens to a star when all (or almost all) hydrogen in its core burns out? When hydrogen fuel fits to the end nucleus stars starts shrink, a his temperature swiftly is growing. AT result formed very dense and hot region, consisting from helium With small impurity more heavy elements. Gas in such a state is called degenerate. Nuclear reactions in the central part of the nucleus practically stop, but enough actively continue leak on the his periphery. The star begins to swell rapidly, swell by leaps and bounds, and its size and luminosity much increase. Star coming off With main sequences and is turning in red giant With temperature surfaces near 3 thousand degrees Kelvin.

However in central areas swollen stars helium continues transform in carbon and oxygen up to before most heavy elements. What will happen when the helium fuel also runs out, like hydrogen in the previous stage? Further move events depends from initial masses stars. If a she is was small like our sun, external layers dumped, forming planetary nebula (an expanding cloud of gas), in the center of which the already familiar to us lights up white dwarf - hot star size about With earth and With weight order massesSun. Medium matter density white dwarf is $10^6 g/cm^3$.

White dwarfs - very curious objects. Representing yourself on essence affairs, dead star (thermonuclear reactions a long time ago got off on the No), they continue radiate, and gravitational contraction is nevertheless unable to overcome the counteractingto him high pressure. Straightaway same arises question: where this is pressure is taken if temperature domestic regions stars relatively low (really So), a thermonuclear reactions ordered to live long? Paradoxical laws are "guilty" of everything quantum mechanics. Under action gravity substance white dwarf compacted so, what atomic nuclei literally squeeze in inside electronic shells neighboring atoms. Electrons lose intimate connection co their relatives atoms and begin to travel freely in interatomic voids throughout the space of the star, then time how naked nuclei form sustainable tough system - some similarity crystal lattice. This state is called a degenerate electron gas, and although white dwarf continues cool down, average speed electrons decrease not thinks. According to the laws of quantum mechanics, the closer the electrons are to each other, the their velocities should differ more strongly, from which it follows that most of the electrons will be move very fast. Let's listen physicists:

...

Such quantum mechanical motion is in no way related to the temperature of the substance, it creates pressure, called pressure degenerate electronic gas. At whites dwarfs exactly this strength balances strength them own gravity.

So the way white dwarfs how would "ripen" inside red giants and present yourself final stage evolution majority stars. it dead, gradually cooling worlds, inside which all hydrogen has burned out, and nuclear reactions stopped. By the way, in the distant future, such an unenviable fate will befall our Sun. According to calculations, in about 5-6 billion years it will burn the entire hydrogen and turn into a red giant, increasing its luminosity hundreds of times, and the radius - in tens. It is curious that HG Wells predicted a similar evolution of our luminary in novel "The Time Machine" If you, the reader, remember, its a time traveler saw in the distant future a huge crimson Sun in half the sky, hanging over the desert by sea. frankly saying wells a little bluffed because the swollen Sun was supposed to heat the surface of the Earth to several hundred degrees Celsius, so that the time traveler would be roasted alive along with his clumsy machine. But let's not cling to the classics on trifles. The Sun will live in the red giant stage several hundreds million years, a after throw off shell and will turn in white dwarf.

And how will a more massive star behave after the depletion of helium? If its initial the mass was more than 8 - 10 solar masses, in the center of the star an onion-shaped a core made up of heavy elements surrounded by layers of lighter ones. To some moment, such a core loses stability and begins to shrink catastrophically. This phenomenon called gravitational collapse. Depending on the mass of the nucleus, its central part or is turning in superdense an object - neutron star, or collapses
"to the stop", forming a black hole. The monstrous gravitational energy that is released during compression, tears off the shell and the outer part of the nucleus, throwing them out with a high speed. going on grandiose explosion, accompanied birth supernova stars. Us not known space cataclysms more scale, how outbreaks supernovae; in flow some time such star shines brighter whole galaxies. Gradually dropped gas shell cool down and slow down (in interstellar there is a lot of rarefied gas in space), and over time it will form a gas-dust cloud, in in which the specific gravity of heavy elements will be very noticeable. This is explained by the fact that in during its short but turbulent life, the massive star managed to accumulate many heavy elements up to gland, some some of which flew into interstellar space in time explosion. When gas-dust cloud will start condense under action gravity strength, inside him maybe flare up new star. Similar stars, born on the ruins former received call stars second generations, and our Sun, looks like times refers to number just such stars.

Thus, there is some continuity in nature: massive stars first generations are dying enriching interstellar space heavy elements, which serve as building material for second-generation stars. All chemical elements heavier helium formed in stellar bowels in progress thermonuclear synthesis, a

The heaviest elements were created in supernova explosions. The earth has an iron core which accounts for about a third of its mass, so you can roughly estimate which amount gland spat out prehistoric supernova 5 billion years to that back. Everything that surrounds us on Earth, and the Earth itself, is stellar matter inherited us a legacy. It can be said that nuclear reactions in the interior of stars are the main reason diversity of the environment. In the distant past in the universe of heavy elements It was much less, how now, about how testify data supervisory astronomy. Spectroscopic research showed what stellar public strongly different on his chemical composition. For example, hot massive stars, concentrated in the galactic plane, several tens of times richer in heavy elements, how stars ball clusters, lying near center Galaxies.

Flash supernova - very rare phenomenon. Per last thousand years in our Galaxy broke out Total three supernovae - in 1054 year, in 1572 year and in 1604 year. The supernova of 1572, which broke out in the constellation Cassiopeia, was observed by a Danish astronomer Quiet Brahe. AT maximum period she shone with her brilliance brighter Venus. Supernova 1604 of the year yielded in brightness star Quiet Brahe, but all same and she is in maximum shine competed with Jupiter. It lit up in the constellation Ophiuchus and was observed by Johannes Kepler and Galileo Galileo. As for the supernova of 1054, references to it have been preserved in Chinese chronicles, from which follows, what she is was visible even afternoon, a in maximum shine repeatedly outnumbered Venus. Today counts, what Crab nebula in the constellation of Taurus and the pulsar in it (a rapidly rotating neutron star) are the remnants of the supernova of 1054. The Crab Nebula is a cloud of swirling gas, pierced by torn threads - although slowly, but quite distinctly creeps along sky. It seemed would, nothing special but because the distance before this nebulae exceeds 4 thousand light years, which means that the speed of expansion of its gases reaches 1500 kilometers in give me a sec. Between topics speed conventional gas nebulae in our Galaxy not exceeds 20–30 kilometers in give me a sec. Only monstrous on strength explosion could inform the mass gas so high speed.

Although outbreaks supernovae - phenomenon very rare, on measure improvement astronomical observation techniques began to detect them more and more often. galaxies there are dozens billion and somewhere supernova necessarily flare up. BUT because the in maximum his shine they able outshine galaxy, in which lit up, they can be seen at such distances as are only accessible to modern telescopes. For example, supernova S Andromedae, which exploded in this galaxy in 1885, had absolute stellar value minus 19, from what follows, what her luminosity in for a short time, 10 billion times the luminosity of the Sun. Her even could be seen with the naked eye as a very faint asterisk of the 6th magnitude, but nebula Andromedae separate from our galaxies nearly 2 With half million light years. Today, dozens of supernovae are being discovered in other galaxies in year.

Although all supernova explosions represent the final stage of a star's life, astronomers distinguish several types of them depending on the nature of the spectrum and luminosity. There are usually two types of these rare stars. Type I supernovae - old and not so old massive stars that flare in both elliptical and spiral galaxies. Power radiation supernovae this type especially great. supernovae II type are associated with young massive stars that quickly "ran through" their evolutionary path. They are found in the arms of spiral galaxies, where processes continue to take place. starburst, a in elliptical galaxies they not flare up never.

From supernovae should differ ordinary new stars. They are flare up relatively often (about 100 flares per year in our Galaxy), and the radiation power these stars are thousands and tens of thousands less. Without exception, all new ones are cramped double systems, how rule consisting from white dwarf and normal stars.

The initiator of the explosion is usually a white dwarf, a star burned to the ground, from which only the ashes of long-terminated thermonuclear reactions remained. Due to the proximity between components double systems substance superficial layers satellite overflows on the white dwarf, and when his accumulates a lot of, thermonuclear reactions may ignite again. The process has a flash character and resembles the explosion of a giant hydrogen bombs. Over the course of several hours or days, the star reaches its maximum brightness, and then, for many months or even years, it slowly fades away. The mass of the dropped shell is always much less masses most stars, So what she is not collapses at explosion, how supernova, a remains in intact and safety. Received count, what new lose 1/100000 his masses, whereas in supernovae Type I this indicator fluctuates within from 1/10 to 9/10 as well in supernovae II type - from 1/100 to 1/10. After a certain time, a new star can flare up again (sometimes this happens after a few decades). supernovae stars re not never ignite.

So, after catastrophic explosion massive supernova remains tiny a clot of monstrous density - the so-called neutron star. If the filling is white dwarf represents yourself degenerate electronic gas, then in neutron star there are no free electrons. Its mass is so great that the pressure of the electron gas is not forces resist growing gravitational compression. figuratively saying electrons "pressed" in protons, in result what protons turn in neutrons. Per except for the outer layers of a neutron star (crust), its substance consists mainly of neutrons and very small quantities protons and electrons. Pressure in center neutron star reaches such large values that it can exceed several times density atomic kernels. Of course atomic nucleus too built from protons and neutrons, but there only nuclear forces act on them, and in the case of a neutron star, to he adds the heaviest gravity press. We can say that a neutron star represents a continuous atomic nucleus.

To any visually imagine monstrous tightness bowels neutron stars, remember that the size of an atom is on average 10^{-8} cm, and the size of the atomic nucleus is 10^{-13} cm. So the way nucleus less atom in in general in 100 thousand once, a because the almost all the mass of an atom is concentrated in the nucleus, ordinary matter consists of almost emptiness. For comparison: on the segment between the Earth and the Sun, a little more than 100 solar diameters and almost 12 thousand diameters of the Earth, while between the atomic core and nearest electronic shell (orbit) without labor will accommodate 100 thousand nuclear nuclei. If a we let's squeeze nuclei back to back friend to friend, density substances will increase 10^{15} times and will exceed density atomic nucleus. Density neutron stars is estimated at 5×10^{15} g/cm3, which, by the way, is several billion tons. At weight order two solar masses like an object will be perfect tiny - 10–15 kilometers in diameter.

The structure of a neutron star is very complex and poorly understood. How the substance behaves at densities superior nuclear can only guess. Suggested several models describing the structure of neutron stars, but they all end up in one or another hypothetical degrees. Experts agree on only one thing: a neutron star has a layered structure. The surface layer is a plasma that captures incoming from space relativistic particle, which are moving on spirals along magnetic power lines and intensively radiate in x-ray range. Further goes layer, having a crystalline structure, followed by a layer of heavy nuclei, neutrons and electrons. Even deeper are densely packed neutrons, and in the very center located nucleus from quark-gluon plasma. By direction from surfaces to center density increases from 4.3×10^{11} g/cm3 up to 1.2×10^{15} g/cm3.

A typical neutron star model is a layered onion: the outer bark from electrons and cores, internal bark (superfluid neutrons, nuclei With excess neutrons and electrons), external nucleus (superfluid neutrons, superconducting protons, normal electrons) and the inner core, near which there is a big question mark. By some data, neutron matter maybe there turn into in quark. How known neutrons and protons consist from quark triplets. At not very high quark densities are easily held inside the neutron by the energy of the strong interaction, but in the center of a neutron star, where the density goes off scale, they get the opportunity permeate in neighboring particle, then there is start free travel inside superdense area. Quark triplets fall apart, and then such matter follows consider how quark gas or liquid. By calculations theorists Besides conventional and- and d-quarks (upper and lower, from which nucleons are built - protons and neutrons) in such gas are found in big quantity So called s quarks (weird) which are part of the heavy particles - hyperons. Therefore, such quark stars called "weird". (About subnuclear particles, including quarks and gluons, detail described in chapter "Bricks universe.")

So, according to some models, an ordinary neutron is first born star, and after the matter in its depths makes the transition to the quark state, it evolves in quark star. However, complete clarity in these issues no.

Of course discover neutron star through optical observations impossible. Nuclear reactions do not take place inside them, so there is no radiation either. In addition, the surface area of a neutron star is so small that its apparent brilliance will be completely negligible. But if it is included in a binary system, then the nature of the movement of an ordinary star can give out the presence of an invisible neighbor. However the discovery came, as often happens, from a completely different, unexpected side. In the second half of the past century managed register powerful sources radio emission, whose intensity changed periodically over time. In 1967 Jocelyn Bell, graduate student English radio astronomer Anthony Hewish, by chance discovered absolutely unusual radio source, which the radiated in impulsive mode strictly periodically - every 1.33 seconds. After a short time, three more sources were found with such same short intervals. When version about artificial origin signals fell away (at first started talking about extraterrestrial civilizations and even arose small panic), remained the only one option - natural origin radio pulses. Mysterious radio sources got title pulsars and enough soon were identified With fast rotating neutron stars.

If a take star With parameters our sun (diameter near 1.4 million kilometers and a period of revolution around the axis of 25 days) and compress its substance in a volume with with a radius of about 10 kilometers, then the equatorial velocity, subject to conservation of mass monstrous increase - about 100 thousand times. And the rotation period is billions of times decreases to a thousandth of a second. True, the pulsar found by Bell had period noticeably more, but all equals this is very small value, absolutely atypical for celestial bodies. By the way, the pulsar in the Crab Nebula makes 30 revolutions per second, which is already very close to the calculated value, and the pulsar in the constellation Chanterelles has a period of 0.00155 seconds. It is clear that only such bodies, the linear dimensions of which are measured in tens of kilometers. And if so, then before us not what other than neutron stars.

With a record short period of impulses, we figured it out. It remains to find out where such a powerful radio emission is taken. The top layer of a neutron star is plasma, permeated powerful magnetic field. Charged particles move along power lines and in end ends turn out in areas magnetic poles, where thrown away narrowly focused bundles particles With high energy - So called jets (from the English jet - "jet"). The rapid rotation of the star gives the departing particles additional energy. It follows from the calculations that the compression of the star leads to increase in its magnetic field, therefore, knowing its average value for ordinary stars, we can calculate, what it will be at neutron stars. Magnetic field will increase in 1012 times and will be a colossal value of 108-109 Tesla. Well, since the magnetic pole is not required lie on the axis of rotation (the geographic pole of the Earth also does not coincide with the magnetic one) jet will describe a cone. We will see the pulsar at the moment when it "looks" directly at Earth. AT following instant he "turned away" a then cycle repeats again.

Subsequently Besides radio pulsars were discovered x-ray pulsars, a also sources powerful flow gamma radiation (MPG sources) With toy same most strict frequency. X-ray pulsars are components of close binaries systems. Substance neighboring stars overflows on the his surface under action forces gravity (this phenomenon is called accretion), from which the departing photons. However radiate in x-ray range may and single neutron stars. More recently, in the 90s of the last century, seven radio-quiet neutron stars With extreme big attitude x-ray flow to optical. First assumed what in everyone guilty mechanism accretions: although at lonely neutron stars No brother, she is maybe grab interstellar gas, in as a result of which its surface is heated to a million degrees and begins to radiate in x-ray range. However on row reasons this hypothesis not confirmed. Neutron stars born very hot (temperature surfaces is about a billion degrees), and then gradually cool down, but even after hundreds of thousands of years after birth, her temperature can exceed a million degrees. Therefore, more likely Total, we see seven young and hot neutron stars. All they located relatively near from Earth (about 120 parsec), from what can to conclude, whatThe solar system is currently passing through a region of recent star formation. (So called belt Gould).

So, at the end of its life, the star sheds its gas envelope, and its core begins to shrink rapidly. If its mass was less than 1.4 solar masses, the gravitational the collapse will stop at the white dwarf stage. If the mass of the nucleus is in the range of 1.4–3.0solar mass, it will collapse into a neutron star. If the core is even more massive (more three masses sun), arise failure in unknown - mysterious an object entitled
"black hole". critical value in 1.4 masses sun received call limit Chandrasekara, on name Indian theoretical physics, calculated this parameter.

Under black hole should understand region spacetime, fully closed for external observer. From under gravity covers, forever and ever slammed crushed star, no signal can get out, including including and Ray Sveta. Path inside black holes - road in one the end: any subject, fallen into its incomprehensible abyss, disappears without a trace. So the black hole - a very apt term, reflecting the very essence of this unintelligible object. Eternal repose light quantums on the bottom gravity graves explained relatively simply. The more massive the body, the more energy must be expended to break away from it. surfaces. To break the fetters of gravity (depart from Earth orbit), space ship must develop speed 11.2 kilometers in give me a sec. This magnitude is called the second cosmic velocity, or escape velocity. On the surface of the sun it will be 700 kilometers per second, but the escape velocity for a black hole is speed light, therefore leave her inside nothing can.

It may seem strange to the untrained reader, which is not so crazy heavy an object (over three solar masses) forever and ever stops light rays. Why in such case massive stars easily radiate light? However a business here not so much in the mass as such, but in the volume in which this mass is placed. If we become compress earth, carefully keeping her complete weight, then saw would, what second space speed steadily growing, although weight planets not is changing. When radiusThe earth will decrease to 9 mm, and the density of its matter will increase to 1027 g / cm3 (by 13 orders of magnitude more density atomic kernels), speed running away on the her surfaces equals co the speed of light. After that, the press can be safely put aside. According to the general theories relativity, Earth With this moment will start irresistibly collapse on one's own, bye on the her place not formed microscopic black hole.

The term "black hole" was coined by the American physicist John Wheeler in 1969. year, although performance about exclusively massive bodies not emitting on this cause of light, arose much earlier - at the end of the 18th century. In 1783 the Cambridge teacher and amateur astronomer John Michel suggested what in nature must exist compact and heavy heavenly body, on the surfaces which speed escape will exceed the speed of light. The numerical value of the radius at which the speed of light equalizes with the second cosmic velocity, it is easy to calculate for any body if its weight is known. This value is commonly called the gravitational radius (r_g), and it easily calculated by the formula $r_g = 2GM/c^2$, where G is the gravitational constant, and - the speed of light. In the case of the earth, as mentioned above, gravitational radius will be 9 mm, for the Sun it will be equal to 3 kilometers, and very massive bodies (on the order of several billion masses sun) will have gravitational radius, superior dimensions solar systems. Similar kind supermassive black holes, how consider astrophysicists, meet in nuclei spiral galaxies.

A black hole is a strange object. If you look into her dark insides, there will not be found even the slightest signs of matter, but only complete emptiness right up to the very center, where sits So called singularity - dimensionless dot With endlessly big density, in which focused all weight black holes. On the this fact indirectly the above formula also indicates: if the black hole were uniformly filled substance, then the volume would be proportional to the mass, and not the radius. However, especially sensitive people who avoid infinity in any her hypostases, can count the core of a black hole by some kind of quantum of space with a diameter of 10^{-33}cm (so-called Planck length). Then the density of the unimaginably squeezed matter will be express yourself extremely big, but after all final number - 10^{-93}g/cm^3(Planck density), that's why matter, swallowed black hole not shrinks to a point with zero dimension, but occupies such a tiny volume (on the order of 10^{-99}cm^3), which is somehow awkward to call volume. About all these difficult things detail tells in "perinatal" chapters, dedicated birth our Universe ("Comprehensive inflation", "AND dark came" "Imaginary time Stephen Hawking").

If around a black hole at a distance of its gravitational radius to build some conditional sphere, covering the singularity from all sides, we get a physical border this amazing object, called horizon events, or sphere Schwarzschild, on name famous German astrophysics. All, what located under horizon events, fundamentally unavailable, for in framework general theories relativity, time is closely related to space and directly depends on strength gravity. Important emphasize, what horizon events by no means not is real the surface of a shriveled object, but is a conditional boundary, forever separating our simple and understandable world from the offal of a black hole, where everything is violated famous physical laws.

Since the course of time depends on gravity (the more massive the body, the slower flowing time on the his surfaces With points vision remote observer), on measure approaching the event horizon, the clock will continuously slow down until the hands not freeze in complete immobility. On the horizon events time stops at all, but only from the point of view of an external observer. As physicists say, anyone can small time interval on the event horizon corresponds to an arbitrarily long time span at a point at infinity. If the black hole is not rotating, the radius event horizon is exactly equal to its gravitational radius, but for rotating black holes he less gravitational radius. Perhaps, costs more once remind, what horizon events - this is his kind semipermeable membrane, which admits moving material tel only in one and only direction - to center black holes, where reign unknown us laws quantum gravity. If a we let's climbunder horizon, to to inquire how looks singularity, return back will no longer be possible. Moreover, to tell about what exactly we saw there is also not it will turn out for no physical signal not will be able get out from under invisible but quite real covers. Although information - concept perfect, but she is certainly implies the presence of a material carrier, and he will be buried forever below the horizon. The Singularity with all its mysteries is securely hidden from the outside and stubbornly not given in arms. God is not endures naked singularity, joking physics.

Almost every book on cosmology gives an example of travelers, trapped in the vicinity of a black hole. We, too, will not be original and move onalong the beaten track. So, let's imagine that in orbit around a black hole is circling the spacecraft from which the descent module with the astronaut on board is separated. The brave explorer set out to penetrate the event horizon in order to across explore offal black holes. What will see his satellites, remaining on the board ship, and what he will see myself? The crew of the spacecraft surprised to find that as it approaches the event horizon, the module's velocity drops to almost zero. With each in a second it moves slower and slower, barely crawling, like a sleepy fly, close hovering over the horizon, but cannot cross it in any way. The crew of the spacecraft so you will never see how the module dives under the horizon, because for this need spend infinitely time.

Suppose what astronaut every minute sends signal their satellites remaining on board the ship. At first, the signals regularly follow each other, but with some moment intervals between them start irresistibly grow. Module how glued hangs close to the horizon, and the signals come less and less. And suddenly like a knife cut off - complete silence. The companions of our brave pioneer may live before deep gray hair but So and not will hear next signal. To hisregister, them had to would wait kiss eternity. BUT between topics astronaut in reentry module continues properly, every minute send signal per signal...

Now let's move on the board module and let's see on the happening eyes astronaut. He effortlessly crosses the event horizon and plunges into the unknown. the interior of a black hole. True, he will not have to triumph for long, because the tidal forces will first stretch his body in the manner of spaghetti, and then crumble into small vermicelli. essence tidal effect is in volume, what gravitational strength With different intensity affect on the diametrically opposite points extended object. On the Earth we this not notice because two meters difference on height between the crown and the heels is too small for the relatively weak gravitational could show up. Another thing is a black hole with its monstrous gravity. two meters below horizon events - colossal distance, that's why human body will be inevitably torn to pieces. However, such a pronounced tidal effect is observed only for small black holes. If our astronaut dives under the event horizon supermassive black hole (of the order of millions and billions of solar masses), with it absolutely nothing will happen. He will be able to fully enjoy the opened before him spectacle and their own eyes will see finally notorious singularity, only here tell about this extravaganzas will be no one. Being under event horizon, there is no way to send a signal outside. the fate of our traveler sad: inside black holes all roads lead in Rome, then I mean to her center, that's why early or late tidal strength grow up so, what to him bad luck.

digest similar things not easy. Robust meaning starts immediately protest, when speech comes in about such objects, how black holes. But what such robust meaning? Intelligence smart monkey, which grew in terrestrial biological niche. To Unfortunately, the real world, the world of monstrous temperatures and unimaginable pressures, is not It has intersections With our worldly experience. However, the deceased domestic astrophysics AND. FROM. Shklovsky in his time managed come up with good analogy allowing more or less visually imagine unimaginable.

...

interesting analogy can spend between transition from life to of death for everyone individual and passing any object through Schwarzschild radius inside some black holes. Like to that how With points vision *external* observer last thing event *never not will happen* With points vision individual or rather, his "I", his own death is unimaginable and in this sense also never will happen. It should be noted that in this analogy the concepts of "internal" and "external" as would are changing places. If a in "astronomical" case world With his spatio-temporal relations is determined *outside* surrounding black holes Schwarzschild spheres, then in "psychobiological" real consciousness individual is *inside* him, being inextricably linked with his "I". The author would be happy if professional philosophers developed this analogy ‹…›

...

Maybe to be, this is clarified would some before now since unresolved Problems the relationship between the individual and the environment of which he is a part. In the meantime how not recall poetry Selvinsky, written years thirty back, in which develops close idea:

> *Think how this is good…*
> *We only live! Nowhere and never we*
> *will not see our own corpse. We we*
> *only die for others*
> *but for myself we die not Can.*

Let's return to our space travelers. So the crew aboard the ship sees a module sewn to a black hole, because the passage of time on the event horizon, with points vision distant observer, slows down endlessly (can to tell, what time stopped). Time stretched out like a perfect rubber cord from a school textbook physics, and not in forces overflow from one moments to to another. time more not exists, only one infinitely long second remains. As the poet said: "And half-asleep hands are too lazy / tossing and turning on the dial, / And the day lasts longer than a century, / And not ends hug." One in a word, look perfectly not on the what.

But the passenger of the module, if he looks out the window, on the contrary, will see an extremely interesting. FROM ease slipping through under horizon events and absolutely this not noticing, he will begin to rapidly plunge into the depths of the black hole. Kantian stellar sky above the head will cringe, until will become literally in sheepskin, a passenger it seems what he went down on the bottom gigantic well. Monstrous gravity twists space ever tighter, and time outside the black hole, little by little, begins speed up your run. And now it is already flying at a gallop, and years, centuries and millennia flash by as in kaleidoscope. The descent into the Maelstrom continues, a terrible pinhole into nothingness all nearer and closer and time has turned into raging vortex.

In a matter of fractions of a second on his watch, the traveler will see the distant future Universe. He will see how burns down Earth in chromosphere swollen sun, like not It was

5 billion years, as the Sun itself sheds its gas coat and turns into a white dwarf, as stars fade and die. The entire history of the universe will fit into a vanishing a small moment, and the arrow of time, which until recently left for eternity, will shrink to a point. All upcoming events end times will happen at once and suddenly.

However, the oddities of black holes don't stop there. Time inside a black hole maybe throw out such knee, what only hold on. For example, spatial and temporary coordinates may change places. If a would passenger module, With points view of the crew of the spacecraft, by some miracle managed to penetrate under the horizon events (for example, the crew waits for this event indefinitely), then for the external observer (in this case, this is the crew of the ship), the passenger of the module would no longer move in space, but in time. Astronaut inside a non-rotating black hole will see not only another the universe causally not related With our but and his own future.

If the black hole rotates (it is very difficult to imagine a point object with zero dimension, which the spinning around own axes), she is acquires more more unusual properties. AT this case radius horizon events becomes less gravitational radius, and sphere Schwarzschild turns out inside So called ergosphere, which represents yourself vortex gravitational field. All body, her captured, doomed on the relentless traffic. If a astronaut dive under horizon events rotating black holes, he will be able see not one a lots of others universes causally unrelated to ours. Moreover, many physicists, not without reason, It is believed that at the bottom of this jet-black whirlpool, a corridor opens up leading to the so-called white hole - a black hole turned inside out. Substance drawn in under the event horizon by an insatiable black hole, immediately ejected into a parallel universe. AND. D. Novikov, supervisor Center theoretical astrophysicists at University of Copenhagen writes: "Everything that falls into a black hole ends up in another Universe... more before will be absorbed black hole."

Such wormholes (wormholes in English), connecting between yourself isolated worlds, causally unrelated to each other, scientists agreed to call molehills burrows. If a compare black hole With hell, With last in circles Dante's hell, then the exit from it can be likened to Eden, or at least to purgatory. However potential traveler, slipped on mole burrow in another the universe not will be able share impressions about seen, because the tunnel, leading in white hole - road With unilateral movement. Return back to him notallow laws physics.

Necessary Mark, what all without exceptions black holes indistinguishable how twin brothers (or sisters). They all have the same face. Whatever the initial conditions their formation, diversity fades without a trace, and the output is always an automaton Kalashnikov. Any black hole is characterized by only three parameters - mass, angular momentum (spin) and electric charge, and everything that falls into it, too loses individual characteristics.

If a more 20–30 years to that back black holes were considered graceful theoretical speculation a in them real existence It was permissible doubt, then today 99% astrophysicists convinced that black holes already discovered, although the Nobel premium for their discovery has yet to be awarded to anyone. The easiest way to observe black holes is in close binary systems consisting of a normal optical star and an invisible component, on the surface of which the matter of the neighboring star flows. At the same time around the black hole formed So called accretionary disk, similar on the spinning whirlpool. The matter falls into the black hole in a narrowing spiral, and the speed of its movement in internal parts of the accretion disk reaches huge values close to the velocity Sveta. Gas warming up before hundreds million degrees, and black hole starts powerfully emit in the x-ray range. The main release of energy occurs long before Togo, how substance disappear under horizon events, that's why x-ray radiation may be registered by an external observer. By a number of parameters it is noticeable differs from X-ray jets (ejections) of neutron stars, so here it is quite available differential diagnosis. To present time discovered over twenty x-ray objects in low-mass double systems, which considered candidates for black holes. If we add supermassive black holes to this list in nuclei galaxies, then them number exceed three hundreds.

All black holes can be divided into three types: 1) black holes with a mass of 3 to 50 solar masses, representing yourself product evolution massive stars; 2) supermassive black holes in the cores of galaxies reaching 106–109 solar masses; 3) so called primordial black holes, formed in the early stages of the life of the universe. His appearance on the light they obliged local deformations metrics space-time in first moments after Big explosion, long before Togo, how lit up first stars. Because black holes gradually evaporate (mechanism them quantum evaporation was predicted by Stephen Hawking), could survive to this day primary black holes only with weight more 1012kg.

AT conclusion this chapters - small quote from books "Astronomy: century XXI".

...

So, thanks to space research and the commissioning of large terrestrial telescopes new generations open hundreds massive and extremely compact objects, observed properties which very similar on the properties black holes, predicted by Einstein's general theory of relativity. One can hope that ‹…› in the nearest decades will be finally proven Existence black holes in Universe. it will lead to breakthrough in understanding nature space-time and entities gravity.

Something about healthy sense

Try to get me That-FAQ-
Can't-Be! Write down your
name To in haste not forget!

Leonid Filatov

A person who for the first time came into contact with the picture of the world that the modern physics, or with cosmological models of the evolution of our Universe, sometimes experiences real intellectual shock. It begins to seem to him that scientists deliberately they pile absurdity upon absurdity, as if trying to outdo each other, so this painting not fits in in habitual representation about reality. Involuntarily remembered famous statement Nils Bora on about another tricky hypotheses: this idea is certainly crazy, but the whole question is whether it is crazy enough, to to be true. Between topics Bor at all not felt fool a Total only wanted emphasize that undisputed fact, what contemporary physics coming out on the such levels comprehension reality, which completely deprived visibility and not have analogies in everyday worldly experience.

Elusive shadows hide behind the facade of everyday life, eluding everyone and everyone. definitions. When we say that this object is green, this one is red, and that that one is blue, everyone intuitively understands what is at stake. However, in reality no blue no color, yes only strictly certain wavelength electromagnetic radiation. A bee or dragonfly perceives blue in a completely different way, because their the compound eye is arranged differently and is able to see in the ultraviolet range. Them blue and our blue is earth and sky. Dragonfly blue color will surely be much richer shades and semitones, although length waves relevant site spectrum in both cases will remain exactly the same. The subjective picture of the world is very often not has nothing to do with the wrong side of things that are fundamentally inaccessible to ordinary perception, which is guided by considerations of common sense. The sense organs are not a golden key and not a magic lockpick, but just a handy tool that helps species to adapt to their environment. Modern physics goes further leaves from visibility, operating categories, which may to be adequately described only on the language strict mathematics. More at all recently atom painted in form miniature solar systems: positively charged nucleus in center in roles a tiny luminary and negatively charged nimble electrons, spinning like planets around kernels. Today we we know what this idyllic picture not It has With nothing to do with reality. First, electrons cannot be located on arbitrary orbits around core, a forced occupy hard fixed levels, which are determined by the energy available to one or another electron. This is partly resembles a ladder: you can jump from step to step as much as you like, but hang between them - sorry, move over! Secondly, electrons are not at all like solid planets-balls, although we say that the electron revolves around the nucleus. Actually neither about what movement in habitual understanding this the words here not maybe to be and speeches: the electron does not spin as wound up, but is in a certain state, which described complex wave function. Other words we we have right talk only only about *probabilities* stay electron in toy or different point.

And not hurry up exclaim, what this not maybe to be. It happens anything and if So called common sense frankly gives in, refusing to separate the wheat from the chaff, this more not occasion, to throwing away in wastebasket puzzling scientific construction.

One can recall an episode from the Strugatsky brothers' story "The Snail on the Slope" when Pepper (one of the main characters) unsuccessfully tries to get an appointment with the director of a certain the mysterious Department of Affairs of the no less mysterious Forest. Kim, Pepper's boss, his consoles and says that everything will work out with time, and when Pepper screams in his hearts that this ridiculous secrecy is already in his throat and he wants to know at least such a little as director looks, then receives an exhaustive answer.

...

– Which? low growth, reddish ‹...›
– BUT Tuzik He speaks, what he lean and wears long hair, because what at him Noone ear.
– it which more Tuzik?
– Driver, I told you. Kim bile
laughed.
– Where chauffeur Tuzik maybe all this is know? Listen, pepper, it is forbidden same to be so gullible.
– Tuzik He speaks, what was at him chauffeur and several once his saw.
– Well and what? lying, probably. I was at him secretary a not saw his neither once.
– Whom?
– Directors. I for a long time was at him secretary bye not defended dissertations.
– And neither once his not saw?
– Well naturally! You imagine what This is true simply?
– Wait, where same you you know, what he reddish and So Further?
Kim shook his head.

– Pepper," he said affectionately. - Darling. Nobody has ever seen a hydrogen atom, but everyone knows that it has one electron shell of certain characteristics and a nucleus, consisting in the simplest case from one proton.

There is a lot of sadness in a lot of knowledge, our wise ancestors said. Why waste appeal to sound meaning? If a some theoretical statement entirely and fully consistent with experimental data, it should be recognized as true, and not dealt with empty scholasticism. A strong and reliable model has been built, and as long as it works, why you more? If it stops working, another one will take its place. Science is not a religion, it does not fascinate the sacramental question "what is truth". Science does not offer definitive solutions, but builds models. But at this not should forget, what any model vague and imperfect; she is neither in who case not reality, a only her imprint, and borovskayamodel atom not at all similar real atom.

And if popularizers from physics talk about the dualism of properties inherent in to the entire population of the microworld, one must always remember that this is nothing more than a figure of speech. It is forbidden to tell, to they strongly made a mistake against truth, because the electron really behaves like a real conjurer, in the blink of an eye changing appearance: then will turn into a wave, otherwise it will demonstrate its corpuscular properties from the heart. Actually it's all our fault suffocating stereotypes that have the most indirect relation. The electron is neither a wave nor a particle, since the reverse side of things going on not under person; electron - just electron, two-faced Janus, behaving the way he was meant to. In some cases, it acts as a particle, and in others - how wave, staying at this incomprehensible thing in yourself With fixed mass, negative charge and half-whole spin.

Albert Einstein's theory of relativity (both special and general) too contradicts our everyday experience. If a you, reader, able visually imagine a curved three-dimensional space, then honor and praise to you, but most people are definitely not ready for such feats. Meanwhile, the curvature of space near massive celestial bodies - an indisputable fact that has been demonstrated more than once experimentally. And the law speed addition in special theories relativity? If a driver "penny" rides co speed 60 kilometers in hour, a cyclist - co speed 30, and both of them are moving in the same direction, then even a student of the initial schools can easily calculate them relative speed friend.

Now imagine a spaceship flying in pursuit of a light beam with speed of 250 thousand kilometers per second. Let me remind you, just in case, that the speed of light inempty space equals 300,000 kilometers per second. Question: What is the speed of light beam runs away from ship? Human co medium education maybe to think what his are being held for a fool, because the answer, it would seem, suggests itself - 50 thousand kilometers in give me a sec. However, it wasn't there! By measuring the speed of a beam of light, we get, no matter how strange, the same 300 thousand kilometers per second. Moreover, the aforementioned space the ship may come close to the light barrier, but the speed of light, measured atits board, still will not change one iota and will still be 300 thousand kilometers in give me a sec.

A business in volume, what speed Sveta in void - magnitude absolute, this is one from fundamental constants. It is even more striking that this speed is distinguished by a strict constancy. From everyday experience, we know that any body moving by inertia,once slowed down, it will not be able to pick up the initial speed. Let's say rifle bullet, breaking through right through inch board, will fly slower. BUT here light leads myself completely different. If you put a glass prism in the path of a light beam, the speed light will decrease, because in glass it is less than in emptiness. However, it only costs a beam of light break free like its speed will again increase abruptly to 300 thousand kilometers in give me a sec. AT void light always distributed by With one and toy same speed, and influence on the her fundamentally impossible.

On the other hand, all bodies with nonzero rest mass can only move at speeds less than the speed of light. And the faster such a body moves, the more increases his weight and topics slower go established on the German watch. Theoretically, it is possible to accelerate an elementary particle, such as a proton, to such a speed, what his weight will exceed mass all our Galaxies. To accept similar statement not easy, but in reality it is. Habitual ideas about nature of things turn out bankrupt at speeds, approaching to speed Sveta.

And one cannot ask why nature acted in this way and not otherwise, like question long away not always correct. Smooth With topics same success can ask, why speed Sveta equals 300 thousands kilometers in give me a sec, a not another size - greater or lesser. One may wonder why nature needed limit speed dissemination signal some marginal size. Why material bodies cannot move with arbitrarily high speed? All this absolutely empty questions, not having rights on the Existence. Why, why... You can crush water in a mortar until you turn blue. By head and cabbage! This is how the world works and remake his nobody more not succeeded what would neither spoke on this about orthodox Marxists.

The law of conservation of energy was formulated almost 300 years ago, but still So far, nothing is known about the mechanisms by which this law works. All processes are running so that energy is conserved. Equally absurd are the arguments about what happened when the world did not exist. It was. Between by the way, this is understood more ancient. Blissful Augustine in his time used to say that the world was created not in time, but along with time, so talk about the existence of anything before the moment "zero" does not make any sense. Whats up say? Headed was a pop, and modern astrophysicists will subscribe to each of his word.

Unfortunately, there are questions that do not have the right to be raised. While science was floundering in diapers and asked nature about simple and familiar phenomena, the answers sounded quite meaningful. Scale human claims was in that time comparable With his own scale. However, the laws of nature change beyond recognition when the forces fields and distances are beyond our daily experience. We had to ask whether matter is a particle or a wave, the answer turned out to be so unexpected that reason refused accept it. We insisted on a tough alternative, but from the point From the point of view of nature, the question in such a formulation was meaningless. Should be once and for all assimilate what Universe created not for the sake of us, we only side product her evolution,and therefore the answers that nature presents us do not have to fit into the kind our heart scheme. Ask too necessary With mind.

The American science fiction writer Robert Sheckley has a wonderful story called simply and co taste - "Loyal question". Some mighty galactic race, a long time ago sunk in non-existence, built unique unit, knowing all on the light. He could answer any question if it was put correctly. Hearing, as you know, the earth fills up, and legions of enthusiasts surf the cosmic expanses without losing hope find the legendary Defendant. Some succeed, and then those who are lucky,rush to ask the wise car question about the Most Important. Someone asks about the crimson, someone - about the law of eighteen, and someone - about life and death, like Stalin's Pasternak, because each people has its own ideas about the nature of things. However, all walkers inevitably fail. Unfortunately, the Respondent is bound by correctly placed questions a such questions require knowledge, which asking not have.Ask explanatory question turns out nearly impossible task. Earthlings too not lucky.

...

defendant introduced himself them white screen in wall. On the them sight, he was
⟨...⟩ extremely simple.

— Highly OK. defendant, - addressed Lingman high weak voice, - what such life?

Voice resounded in them heads.

— Question deprived meaning. Under "life" Asking implies privatephenomenon, explainable only in terms whole.

— Part what the whole is life? - asked Lingman.

— The question in real form not maybe resolve. Asking all moreconsiders "life" subjectively, co his limited points vision.

— Answer same in own terms - said Morran.

— I only answer questions," the Respondent said sadly. It has
come silence.

— Expanding whether Universe? - asked Morran.

— Term "extension" inapplicable to given situations. Asking operatesfalse the concept of the universe.

— You can us to tell though something?

— I I can answer for any right delivered question, touching natureof things.

One word, unlucky stargazers not got lucky. They are judged Yes rowedSo and so, but sense from them efforts It was a little. last try looked So:

...

— What there is death?

— I not can define anthropomorphism.

— Death - anthropomorphism! - exclaimed Morran, and Lingman fast turned around. - Well finally we moved from places.

— real whether anthropomorphism?

— Anthropomorphism can classify experimental: how BUT - falsetruth or in - private truth - in terms private situation.

— What here applicable?

— And then and other.

They did not achieve anything more concrete. For long hours they tormented the Respondent, tormentedmyself, but true eluded all farther and further.

...

Nesolono slurping heroes set sail home. Here how ends story:

One on the planet - not big and not small, a how once suitable size - waitedRespondent. He not maybe help topics who comes to him for even defendant not omnipotent.

Universe? Life? Death? Crimson? Eighteen? Private truth, half-truthscrumbs great question.

And mutters defendant questions myself yourself faithful questions, which nobody not maybeunderstand.

And how understand them?

To right ask question, need know big part response.

If with sin in half we managed to find some patterns of the microcosm and even experimentally check something, this does not mean that we will get answers to all the damn questions. The true nature of things is still not given into the hands, and it is not for nothing that Leo Davidovich Landau tore and metal when he prepared for printing the popular brochure "What is theory of relativity?". "It doesn't climb into any gates," he fumed, turning to his co-author Yuri Borisovich Rumer - two crooks trying to convincesimpleton, that he will sort out the problem for a dime." Of course, Landau was absolutely rights. Analogy and metaphor - things good ones, but and they early or late start slip. At everyone desire we not Can visually imagine space-time foam in the area of Planck lengths or rolled up into the thinnest tubes are extra dimensions, because Homo sapiens is just smart a monkey who managed to master speech and conceptual thinking. Our senses are hard tied to biotope under name "planet Earth", where us raised and nurtured on the over 3 billion years. You can't jump above your head, and therefore the real background world order, remaining secret per family seals, entirely and beside maybe to be shown just mathematically.

World functioning on universal laws, called laws nature, and mathematics acts as a guide to non-human areas of the world. Intelligence, formed in terrestrial biological niche, on the everyone step passes before paradoxes that cannot be bitten, smelled or picked up. For the one who fell into black hole, space takes on the appearance of time, since it cannot return back, just as it is impossible to move backward along the axis of time, that is, into the past. Imagine visually such picture not easy, but maths, how a thread Ariadne, allows you to penetrate into such nooks and crannies of the universe, where the way is ordered for mere mortals. Truth, some scientists claim what understand in similar things so same at ease, as one tastes salty or sour. Actually they are a little cunning: in reality they understand Total only conformity theories and experiencedresults.

Physics With mathematics - this is narrow path above abysses, inaccessible human imagination. Man is so constituted that he longs for final truths, but in science needed restraint. World refuses reply on the questions about his ultimate essence, and we are lost when we learn that the absolute vacuum is not at all empty, a energy maybe to be negative. Between by the way, exactly in this rooted specific difference between faith and knowledge. Faith all knows ahead, she has, how dexterous sharpie, always hidden in sleeve trump card map. BUT the science distinctly conscious his imperfection. Maths can do a lot but far away not all.

Unfortunately, mathematics does not always help, because there is no certainty that the world is mathematical in nature. Of course, this clever code sometimes allows you to get answers to correctly posed questions, but this does not mean that mathematical symbols reveal essence of things. Of course, we are not so naive as to cross out mathematical an approach in principle we only emphasize purely ancillary role mathematics how cognitive guns, helping reach certain goals. O there is no identity between the object of cognition and the instrument of cognition of speech. Stanislav Lem So wrote about it:

...

Mathematics is more like a ladder to climb mountain, although it does not look like this mountain at all. ‹…› From a photograph of a mountain, you can use corresponding scale, determine its height, slope fall and So Further. Stairs can also tell us a lot about the mountain to which she was leaned. However, the question of what grief corresponds rungs stairs, not It has meaning. After all they serve for Togo, to get to the top. Similarly, it is impossible to ask whether this ladder "true". It can only be better or worse as an instrument of achievement. goals.

Gold words. In fact, the point here is that our models, even if they perform well, agree remarkably well with experience, and produce predictable results, may turn out to be just a pale shadow of an incomprehensible reality. And it's even better case. What if someday it turns out that all our models, stuffed with puzzle mathematics, not have smooth account no relations to the world of things? Such unpleasant perspective too should have in mind on the any happening. And although pragmatic aspect of scientific theories from this will not suffer in the least, it will still be up to depths souls it's a shame be aware what humanity never not destined get through to fundamentals of life. This deeply philosophical question was wittily played up by the already familiar us Robert Sheckley.

In his brilliant novel The Exchange of Minds, there is a small chapter devoted to called the Distorted World - unsteady and whimsical inside out boring reality. let's let yourself a few quotes.

...

...so thanks Riemann-Hacke equations was finally mathematically proven theoretical need twister-mann spatial zones logical deformations. This zone received title distorted Mira, although on the himself deed not distorted and the world not is.

And Further:

...

A certain sage once asked: "What will happen if I enter the Distorted World without having preconceived ideas? It is impossible to give an exact answer to this question, but we believe thatby the time the wise man comes out of there, he will have preconceived ideas. Absence beliefs not the most reliable protection.

...

Some consider the highest achievement of the intellect to be the discovery that absolutely everything can be turned inside out and turned into its own opposite. Based such an assumption, you can play many amusing games; but we do not call for his in Distorted World. There all dogmas equally arbitrary including dogma about arbitrariness dogma.

...

Not hope outwit Distorted World. He more, less, longer and shorter, howwe. He is unprovable. He just eat.

...

What is already there does not require proof. All evidence is an attempt at something. become. The proof is true only for itself, it does not testify to anything, Besides availability of evidence a it's nothing not proves.

...

That, what there is, incredible, for all alienated, no need and threatens reason.

...

Maybe, these remarks about Distorted world not have nothing general Withdistorted Peace. But the traveler warned.

Of course, the uncle is joking, but, as you know, in every joke there is always a share of a joke. Worldturned out to be much more complicated than our home-grown ideas about it, and there is not a single word about it. minute should not be forgotten. Of course, the last thing I want is for you, the reader, thought that nature was unknowable. I was just trying to emphasize what needs to be soberly evaluate their capabilities, a not study cheap capping.

bricks universe

> *Praise to that who the first began call cats and catshuman names*
> *Who   gave beetles the names of grinders, gravediggers andlumberjacks,*
> *Who decorated teaspoons with letters and monograms,*
> *Who Greeks divided on the ancient and for just Greeks.*

Nicholas Oleinikov

antique philosophers thought what foundation universe complicated from four the basic elements are earth, air, fire and water. The great Aristotle added to this combinations of the fifth essence - the so-called quintessence, from which essential body. He thought what substance can fraction endlessly, So never and not having reached before toy smallest grains, which already not lends itself furthercrushing. Stubborn atomists disagreed with the luminary of all sciences, insisting that matter is made up of atoms - tiny indivisible particles that are in constant movement (the word "atom" in literal translation from Greek means "indivisible"). This idea supported such outstanding thinkers antiquities, how Democritus, Epicurus and Leucippus, but since ancient science was thoroughly speculative and afraid of experimentlike the devil of incense, there was little sense from these exercises in vainglory. Even when English naturalist John Dalton showed in 1803 that chemicals are always united in certain proportions, the centuries-old dispute between the two schools is stillnot was finally resolved in benefit atomists.

However, in the century before last, the vast majority of scientists no longer doubted corpuscular structure of matter. By the end of the 19th century, when Joseph John Thomson of Trinity College, Cambridge discovered the electron, it became clear that the atom has a complex internal structure and not is elementary brick universe. But whatelectrons and protons (the neutron was discovered only in 1932 by James Chadwick)located in the atom relative to each other, it was not at all clear. Say lord Kelvin thought atom spherical education, on everything volume whom evenly positive charge is distributed, and inside the sphere in static equilibrium are negatively charged electrons. But just a few years later, Rutherford did not left from this model stone on stone.

An experience English physics was relatively simple. He shelled the thinnest golden foil bundle alpha particle, flying co speed twenty thousand kilometers in give me a sec. alpha radiation - this is massive positively charged particle, emitted some nuclides in process radioactive decay. Rutherford occupied question, how much strongly deviate particle, passing through golden foil. Picture turned out very curious. How and should expect, big part alpha particles pierced the foil right through, practically not deviating or deviating by a slight angle of 2-3 degrees. But some particles were deflected much more noticeably - 90 degrees or more, and some few even bounced back at all, as it flies off from wall thrown ball. One got the impression that the atoms of the thinnest film could be serious obstacle on the way swiftly flying massive alpha particles. it seemed completely unbelievable: one could just as well have assumed that the leaf drawing paper able to stop a rifle bullet.

And then Rutherford suddenly dawned. He used the example, as they say, from another opera - he imagined how a comet behaves in the vicinity of the Sun. Caught in a powerful gravitational field our luminaries, she is maybe strongly change trajectory flight, do, for example, coil and retire from sun in himself unexpected direction. FROM on the other hand, the gravitational interaction between the objects of the microworld is so little that it hardly makes sense to take into account. Then maybe inside the atom operate some other strength, for example electromagnetic? alpha particle indeed positively charged, but here's the problem: the atom itself is electrically neutral! BUT what if intraatomic charge distributed uneven? After all comet too interacts not co all solar system, a only With her central link - Sun. And Rutherford guessed that it is consistent to explain the result of the experiment only one way is possible. An atom is made up of a positively charged nucleus and negatively charged electrons that revolve around the nucleus like planets around Sun. And atomic nucleus a lot of less atom in in general (how and Sun much smaller than the solar system), although almost the entire mass of the atom is concentrated just in the atomic nucleus. Therefore, those alpha particles that flew far from the nucleus are almost were influenced by it, but the particles captured by the nucleus deviated very strongly. BUT since the atom, with the exception of the nucleus, is practically empty, the number of perceptibly deviated particles it was very insignificant.

Today we know that the average size of an atom is 10-8 cm, and the size of an atomic nuclei - 10-13cm. Difference on the five orders then there is in 100 thousand once! Charges proton and electrons are opposite in sign and equal in absolute terms, but the mass of a proton exceeds the mass of an electron by 1836 times. In an electrically neutral atom, the number of protons corresponds to the number of electrons, but the protons are collected in a vanishingly small volume (and therethere are also neutrons that outnumber electrons by about the same amount) while the electrons are distributed throughout the atom. So the positive charge and nearly all weight atom extremely concentrated a negative charge sprayed, "smeared" throughout space tiny "solar systems."

Of course, the planetary model atom, proposed by Rutherford in 1911, is not has remained unchanged to this day. The first serious amendments to it were made by Niels Bor and wolfgang pauli, and With flow time atom became all less and less remind solar system. In second half of the past century It revealed, what nucleons atomic nuclei (modern physics thinks what proton and neutron - this is two charge states one and toy same particles - nucleon) at all not initial bricks of the universe, but are built in turn from special subnuclear particles - quarks. This term invented Murray Gell-Mann, theorist from Californian technological institute, borrowed voiced word at James Joyce author abstruse things "Wake on Finnegan." AT 1969 year per study quarks he was honored Nobel premiums.

As we can see, there is almost nothing left of the solar system. And although today we wonderful known what real electron at all not similar on the planet a if his and canWith something compare, then quicker With some blurry cloud, possessing complexproperties, this is not at all not belittles values proposed Rutherford models. Not subject to doubt what myself English scientist in complete measure gave away yourself report in approximate own analogy, although not had concepts neither about principle uncertainty Heisenberg, nor topics more about quarks Gell-Mann.

Nevertheless, Rutherford's model immediately ran into serious difficulties. Since the electron is in constant motion, it essentially represents a moving electric charge that continuously wastes energy, because moving charge must radiate. Consequently, through very a short time exhausted electron, mediocre squandered mine gold stock, must on converging spiral to collapse on the core. In other words, the Rutherford atom is ultimately unstable, he must die in a matter of fractions of a second. Way out of this unpleasant provisions found great Dane Nils Bor, one from creators quantum mechanics.

However, first as should deal with the structure of the atom. In the simplest case atomic nucleus consists from the one and only proton. So arranged for example, atom hydrogen: positively charged proton in center and carrier negative charge an electron in orbit around a proton. In general, the hydrogen atom is electrically is neutral, since plus and minus eventually gives zero (recall that although the electron and proton differ in mass by a factor of 1836, their charges are equal in magnitude). So the structure of the atom simple hydrogen (protium) can be represented graphically as follows: $]H$. Unit bottom left from chemical symbol hydrogen (H) stands for atomic room element, which corresponds to the number of protons in the nucleus (and since the atom is electrically neutral, There are exactly as many electrons in orbits as there are protons. The unit at the top left is mass number reflecting the number of nucleons in the nucleus (that is, protons plus neutrons). AT case ordinary hydrogen, protia, neutrons in core No, that's why atomic room and massive number are equal between yourself.

If we add a neutron to the nucleus of ordinary hydrogen, we get its isotope - deuterium, or heavy hydrogen. Then his formula will be to look like So: $1\ 1H$. Atomic room is still equal to one, because the number of protons in the nucleus has not changed, but the mass number has grown twice because the to proton added not having charge neutron. At hydrogen there is more one isotope - tritium, formula whom sign up next way: $3\ 1H$. It is easy to see that the tritium nucleus contains 2 neutrons and 1 proton (mass number equals three), a here atomic room again same not changed So how proton all more staysin proud loneliness. And protium, and deuterium, and tritium are chemically completely identical and are the same element - hydrogen, because the chemical properties elements connected With valence electrons a them amount in all three cases absolutely the same (number of protons is equal to the number electrons).

So, chemical elements, having same atomic room, but various massive numbers, called isotopes. Or more easier: isotopes - this is nuclei atoms, differing in the number of neutrons, but containing the same number of protons. All threehypostases of hydrogen - protium, deuterium and tritium - will occupy the same cell in Periodic system of elements. Now let's try to apply what we've learned to practice. How known natural Uranus consists from mixtures three isotopes - uranium-238, uranium-235 and uranium-234, and on the share uranium-238 account for more 99 %. Here his formula:

$238\ 92U$. Atomic room uranium-238 expressed number 92, Consequently, in his core contains 92 protons, but the total number of protons and neutrons is 238. To to know, How many in core uranium-238 available neutrons need subtract from more the smaller number: 238 minus 92 equals 146. So, there are almost twice as many neutrons in the nucleus of uranium, than protons. The same applies to its other two isotopes, only the number neutrons in their nuclei will be slightly less. All three isotopes of natural uranium occupy the same cell of the Periodic system of elements and contain 92 protons (their atomic room one and that same). Such overloaded neutrons nuclei very unstable and able to spontaneously disintegrate. This phenomenon is called radioactive decay andaccompanied generation hard radiation (various options radioactive we will not analyze the decay). Incidentally, the tritium nucleus, unlike deuterium and ordinary hydrogen, too unstable because what has an excess neutrons.

Let's get back to atom Rutherford, which the not It has rights on the Existence. How save life electron, which the wastes energy, addressing around atomic kernels? How already said above, solution this Problems found Nils Bor. He postulated what electron situated not on the any arbitrary orbit, a only on theone that lies at some well-defined distance from the nucleus. Moving on to such allowed orbits, electrons do not radiate, and therefore do not lose energy. emission or absorption energy going on at jump electron With orbits on the orbit, and the fact that this energy is quantized, that is, broken up on the his kind portions. Electron seeks take in atom most advantageous in energetically the level where its energy is minimum. The closer the orbit is to nucleus, the lower the energy of the electron located on it. If the orbit closest to the nucleus is already occupied, the electron takes off to a higher orbit, but for this it necessary purchase additional energy, then there is absorb quantum Sveta (electromagnetic radiation). emitting quantum electromagnetic radiation, electron maybe go down a floor below.

Important remember, what all these orbits - how loved ones, So and distant - by no means not arbitrary a present yourself hard fixed energy levels. AT famous sense system electronic shells (or orbits) can liken ordinary stairs. To climb up the stairs, you need to do work, then is to expend some energy. The descent down is incomparably easier, but hanging between stairs is still impossible: at each individual moment of time, the climber must occupy a very specific step. The intraatomic ladder is fixed in the same way hard. An electron that has absorbed a quantum of electromagnetic radiation (recall that this is strictly measured a portion energy), receives possibility take a step on the next step, for his energy increased. measure this energy will be distance between steps. The more energy an electron acquires, the higher it can climb. However, the electron always dreams of returning to the first floor, since this is the most profitable position. He can immediately fall to the initial level, and then the energy of the electromagnetic radiation will be in accuracy is equal to the one which was originally absorbed. BUT here if he get stuck in the middle then his radiation will be give other energy, a Consequently, and length waves. So, energy, acquired or lost electron, determined by distance between steps.

released from atom energy maybe to be registered. BUT because the each chemical element has, so to speak, its own unique set of steps, spectra the radiation of different substances will be highly individual. In other words, each chemical element It has my calling card, what very on the hand astrophysicists. By studying the spectra of distant stars, it is possible to identify thechemical elements.

So, we came to conclusion what borovsky atom not at all not similar on the atom Rutherford. On the other hand, it also has a very indirect relation to the real atom, because that the Bohr atom (the atom that Bohr built, as the famous song parodies famous English poem) is nothing more than a convenient model for understanding essence processes, ongoing in world elementary particles. However before how go to fundamental bricks universe (that is, the aforementioned elementary particles), necessary although would short stay on the principle uncertainty which is the alpha and omega of quantum theory. If the eminent German physicist Max Planck suggested in 1900 that no electromagnetic radiation (visible light, x-ray rays, a also waves any lengths) not maybe be generated With arbitrary intensity, but certainly must be dosed in portions (Planck named these portions quanta), then another famous German, Werner Heisenberg formulated its fundamental principle.

According to the Heisenberg uncertainty principle, it is impossible at the same time accurately measure the coordinates of the particle and its speed. Understand the essence of Heisenberg's reasoning not difficult. If a you want to predict what way will change position and speed particle, you must be able to produce accurate measurements here and now. Absolutely it is obvious that for this you must direct a beam of light at the particle, and the shorter the length waves light beam, topics more precisely to you succeed calculate coordinates particles. However, based on Planck's hypothesis, light cannot be dosed arbitrarily in small portions, for him available some indivisible fragment - one quantum. Clear, what this quantum certainly will contribute disturbance in trajectory particles and unpredictably will change her speed. To achieve greater accuracy in measuring the particle coordinate, you will become shorten the wavelength, and then the energy of the quantum will automatically increase. (Wavelength tied With energy quantum back proportional addiction: how shorter length waves, the higher the energy.) Therefore, the speed will immediately increase. Stephen Hawking, one from pillars contemporary theoretical physics, writes about it So:

...

In other words, the more accurately you try to measure the position of a particle, the less accurate will measurements her speed, and vice versa. Heisenberg showed what uncertainty in position particle, multiplied on the uncertainty in her speed and to its mass, cannot be less than a certain number, which is now called a constant Plank. This number does not depend on the way in which the position or speed is measured. particles, nor on the type of this particle, i.e., the Heisenberg uncertainty principle is fundamental compulsory property our world.

Principle uncertainty It has far-reaching consequences, in volume including and philosophical character. Finally covered herself copper pelvis cheeky dream determinists who, with a blue eye, undertook to predict the future of the universe, if in them disposal will be accurate coordinates all constituents her particles. It became it is clear that the subject and object of knowledge cannot exist without each other and forever tied with one rope.

To touch an object without disturbing it in the least would be possible only for the Lord God, but we mercilessly take it to the dustbin of history, for it is said: one should not multiply the number entities in excess of need (William occam, medieval English philosopher). Occam's approach (or "Occam's razor") was adopted in the 20s of the last century Niels Borom, Werner Heisenberg Erwin Schrödinger and field Dirac, in resulting in classical mechanics gave way to quantum theories, at the forefront which was the principle uncertainty.

Quantum mechanics once and for all crossed out the determinism on which the old physics, and contributed in science inevitable element unpredictability. Wingless and flat uniqueness conceded place for probabilistic approach.

Knowing initial options systems, we already not Can guarantee quite a certain result, but we are talking only about the fact that the system will be in one or otherwise able With some probability. it It was so unusual and marvelous!

Even this heretic and like a revolutionary Albert Einstein, Once Upon a Time with this in hearts declared that God does not play dice. However, most scientists immediately accepted quantum mechanics because the she is gave beautiful agreement With experiment.

From principle uncertainty most direct way follows So called wave-particle duality. Any particle can easily turn around wave, and vice versa: essence of things, how neither strange, escapes from strict formulations. Let's say electromagnetic radiation propagates in the form of fixed portions, or quanta, what earnestly demonstrated Max Planck. However in compliance With Heisenberg's uncertainty principle photons (quanta of electromagnetic radiation) in then same most time lead myself how waves, not having certain provisions in space, but "smeared" over it with some probability distribution. light in given case - by no means not exception; exactly So same lead myself all others particle,which are called elementary.

Physicists are a little cunning when they say that an electron revolves around an atomic nucleus, because in reality about any movement in the usual sense of the word here not maybe to be and speeches: electron not spinning how routine, but located in some certain state, which described complex wave function. In other words, we have the right to speak only about the probability of an electron stayingin one point or another.

Let's finish on the this our short excursion in quantum mechanics and let's move on to consideration elementary particles like those.

If a photon or electron is indisputably elementary, then this cannot be said about the filling atomic nuclei - protons and neutrons because the they have complex internal structure. Both of these particles are quark triplets, that is, they are built from more fundamental bricks - quarks, those most quarks, per opening which Murray Gell-Mann was awarded Nobel premiums. However both everyone on order.

The main properties of all elementary particles without exception are mass, charge and spin. The mass of a particle is a fraction of its total energy, because mass is Total only another her the form. Weight maybe to be transformed in energy, and vice versa; relationship between these two parties one medals easily see in famous Albert Einstein's formula $E = mc^2$, where E – energy, m is the mass, and c is the speed Sveta. Some particles have mass, while others do not. For example, physicists say that the rest massphoton equals zero. it quite simply means what resting photons in naturenot exists. Remains add, what distribution particles by mass not obeys no intelligible patterns.

Electric charge - also a familiar animal. With the charge, the situation is exactly the same the same as with mass: some particles carry it, while others do not. Particles that have no charge are considered electrically neutral. Unlike the mass There are two types of charge positive and negative; charges all elementary particles multiples charge electron, per exception quarks, charge which multiple of 1/3 charge electron.

Spin elementary particles represents yourself some interior moment her rotation and is proportional to Planck's constant. If the particle is not rotating, its spin is zero. From considerations visibility can introduce yourself particles in form smalltops or balls, rotating around his axes, but always should remember, what similar painting purely conditional and not It has With reality nothing general. AT quantum world elementary particles do not have a strictly defined axis of rotation. Particle spin gives us an idea of what it looks like when viewed from different angles. Stephen hawking leads good example on the this account.

...

A particle with spin 0 is like a point: it looks the same from all sides. Particleco back one can compare co arrow: With different parties she is looks differently and takes the same form only after a full rotation of 360°. A particle with spin 2 can be compare with an arrow sharpened on both sides: any of its positions is repeated after half turn (180°). Similarly, a particle with a higher spin returns to the initial state when rotated by an even smaller part of a full turn. It's all quite obvious, but surprisingly different - there are particles that, after complete turns do not take their former form: they need to be completely rotated twice! They say that such particles have spin 1/2.

All known elementary particles can be divided into two groups depending on the magnitude of the spin they carry. If the spin is expressed as an integer (0, 1, 2, etc.), then such particles are called bosons, and if half-integer (1/2, 3/2, 5/2, etc.) - fermions. These titles formed from surnames two famous theoretical physicists Satyendra bose and Enrico Fermi. All matter in the universe is built from fermions - particles with a half-integer spin, and the forces acting between the particles of matter are created by bosons having integer spin. Spin electron is 1/2, that's why he hits in group fermions.

AT dependencies from them relations to strong interaction (about four types fundamental interactions speech at us ahead) fermions, in my turn, are divided into two families. Those fermions that take part in processes with strong interaction, called quarks (protons and neutrons consist from quarks), and all the rest, not participating in strong interactions, are leptons. Electron enters in the family of leptons; in addition to it, five more particles are placed there - an electron neutrino, muon, muon neutrino, tau neutrino and tau lepton. There are also six quarks varieties - i-quark, d-quark, c-quark, s-quark, t quark and b quark. So the way bricks universe, construction blocks matter, which we everywhere we observe are 12 fundamental particles - 6 quarks and 6 leptons.

Among bosons, being carriers fundamental interactions and creating forces acting between particles of matter, photons are best known, 8 varieties gluons, 3 kind heavy vector bosons (W+-boson, W-boson and Z0-boson) and while more not open graviton.

Remains add, what in contemporary theories fields particles act how small-scale waves of the corresponding fields. For example, electromagnetic radiation can be perceived both as a wave (say, in the case of radio waves) and as a particle (hard gamma rays). If a length waves electromagnetic radiation much exceeds dimensions of the device, then it is recorded as a continuous wave, that is, traveling oscillations electric and magnetic fields. Otherwise (at a small wavelength) the device captures light in the form of individual quanta - photons. Then they are no longer talking about the wavelength, but about energy photon. Classical example corpuscular-wave dualism.

fermions, from which built substance of the universe - by no means not indifferent extras on the this holiday life. They are interact between yourself a in roles carriers interactions (or strength, existing between particles substances) act bosons. To create all manifold phenomena, nature it took round account four type interactions - electromagnetic, weak strong (or nuclear) and gravitational. There are strong reasons to believe that the first three types interactions under certain conditions can be combined into one force, and separately they exist only at low energy levels. So far, a model has been built electroweak interaction (electromagnetic + weak), and the carrier particles of this unified force discovered experimentally (three types of heavy vector bosons). Theory, unifying three strength in one (electroweak interaction + strong), called grand unified theory, but the energy level required for this is not available modern accelerators. At even higher energies, all four forces of nature. Such conditions existed in a very young universe, when the world was just flitted out from non-existence.

Let us analyze the four types of fundamental interactions in order. Electrical and magnetic phenomena have general origin and are described in framework electromagnetic interactions, which So or otherwise related With exchange or radiation photons (quanta electromagnetic radiation). First this is showed eminent English physicist James Maxwell back in 1873. Electromagnetic forces operate only between charged particles (of the same name charges repel, dissimilar - attract). Radio, television, cellular communication and many other convenientand useful things unthinkable without phenomenon electromagnetism, because the these strength, based on the confrontation two polar began, able spread on the significant distances. Moreover, the atoms and molecules that make up matter too obliged their existence electromagnetic interaction. Forces electromagnetic attraction hold back electrons inside atoms, forcing them rotate around atomic kernels. AT roles carrier electromagnetic forces speaks a massless particle with spin 1 is a photon (physicists say that the rest mass of a photon is equal tozero).

Interaction between two charged particles (attracted they or repel, in given case roles not plays) represents yourself result exchange a large number of so-called virtual photons. Unlike "real" particles, their virtual sisters are fundamentally unobservable, they cannot be registered with help detector. Let's explain said on the example. Imagine yourself some closed a container with nothing inside - no radiation, no matter. In other words, there contains only vacuum, absolute emptiness. But to make sure the container is really empty, we must illuminate its insides - send a beam of light there. And since light travels at a finite speed, the measurement process will take some time. To say with complete certainty that the container is empty, we can only at that moment, when the light beam returning from the container reaches our detector. At the same time, we have no certainty that the container remained empty *all the time procedures measurements*. Not ruled out, what energy vacuum could hesitate (fluctuate) around zero, giving rise to short-lived ghost particles that die before we can spot them. They emerge from the void and hide in it again so swiftly, what we not Can discover them in principle even if we have most perfect measuring equipment. Such particles received call virtual.

Of course not all photons are virtual. Quanta Sveta, which released in as a result of the transition of an electron from orbit to orbit, are quite real photons. Similarly, when a real photon collides with an atom, an electron can jump over on the more remote from nuclei orbit. AT this case energy photon will be absorbed. So, to summarize: the electromagnetic force acts between all particles, bearing electric charge, a her carriers are virtual photons. BUT because the weight rest photon is equal to zero, electromagnetic interaction maybe be transmitted on the large distances.

Weak interaction answers per some transformation in world elementary particles. Good example forces this type - So called beta decay unstable atomic nuclei, as a result of which the intranuclear neutron turns into a proton, and from nuclei fly out electron and antineutrino. AT weak interaction participate all particles with spin 1/2 (that is, all fermions), and its carriers are heavy vector bosons co back one (W+-boson, W-boson and Z0-boson). Because the vector bosons - extremely massive particles (they heavier proton nearly in 100 once), weak interaction is effective only at ultra-small distances of the order of 10^{-16}–10^{-17} cm. How It has already been said that the weak interaction has been combined with the electromagnetic one. It was done in the standard Weinberg–Salam model, which is detailed in chapter "And Darkness Came". The weak interaction is most closely related to thermonuclear reactions, during which hydrogen in the stellar interior turns into helium, and also to some others processes, accompanying person evolution stars different types.

The strong (or nuclear) force keeps quarks inside nucleons, and protons and neutrons - inside the atomic nucleus, overcoming the forces of Coulomb repulsion (protons have eponymous charge). How we remember exists six varieties (or flavors) quarks - i-quark, d-quark, c-quark, s-quark, t quark and b quark. Them titles educated from English words up - "up", down - "way down", charm - "the charm", strange - "strange", truth - "truthful" and beautiful - "beautiful". Apparently physicists tiredLatin and Greek, and they decided name fundamental bricks top, lower,enchanted strange truthful and beautiful particles. Protons and neutronspresent yourself quark triplets, but in them compound are included only quarks twofragrances - und. Proton built from two u-quarks and one d-quark, a neutron - from twod-quarks and one u-quark. BUT because the d quark a little heavier u-quark, neutron a littleheavier proton. Difference in them charges (proton charged positively, a neutron charge notIt has) too explained features internal buildings, So how quarks bearfractional electric charge (2/3 and -1/3). So the way from three quarks, two from whichhave charge a plus 2/3, a one - minus 1/3, it turns out proton With charge +1. BUT neutronconsists from one quark with charge 2/3 and two with charge minus 1/3, so as a resultcomes out zero. From quarks others types (weird, enchanted, b and t) too can buildparticles, but they are unstable and quickly decay into protons and neutrons.Except Togo, each quark maybe be in three various states, which received call color (red, yellow and green). Of course in realityno colors at quarks No, this is simply comfortable generally accepted designations them properties.Elementary particles consist from quarks different colors, but always in such combinations,to in result turned out colorless particle. For example, triplet "red + green + blue" will turn out to be a proton or a neutron. Closely related to the presence of color in quarks the phenomenon of the so-called confinement of quarks ("non-ejection", "retention" in translation from English). The fact is that quarks never occur in isolation, but exist in close cooperation friend With friend, in form already acquaintances us quark triplets. So far, no one has been able to detect a single quark. If the quark wanted stand apart and live on one's own, he instantly has gained would color, what forbidden conditions tasks: confinement obliges them be held in colorless combinations. However, at very high energies, the strong interaction noticeably weakens, and then quarks begin to behave almost like free particles. Such a quark-gluon plasma existed on the early stages our life Universe.

Quarks held in triplets per check carrier particles strong interactions - gluons (from the English glue - "glue", "glue"), which stick them together between themselves. Gluons have zero mass and a spin of one. Unlike all other types of interactions, nuclear forces do not weaken as quarks move away from each other from friend, a against, are growing. Gluons can liken tight rubber bands connecting quarks to each other. As long as they are side by side, the elastic bands hang freely, allowing quarks feel relatively at ease. But they should try to move away from each other, as the rubber bands immediately stretch and return the mischievous people to their original position. Nuclear strength effective only on the very small distances order 10^{-13}– 10^{-15} centimeters.

It remains for us to consider the fourth type of fundamental forces - gravity, which wears universal character and makes body be attracted friend to friend. gravitational interaction - most weak from all: strength electromagnetic repulsion exceeds constricting strength gravity about in 1043 times. However weakness gravitational interactions With overwhelmed bathe huge dimensions celestial bodies, consisting of an astronomical number of particles, so the forces of gravity between planets or stars may to give very big size. Except Togo, ifelectromagnetic forces act only on charged objects, then gravity exerts influence on the all without exceptions body and particles our universe, possessing mass.

carrier gravitational interactions is bye more not open graviton particle, which should have zero rest mass and spin equal to two. Like electromagnetism, gravitational interaction is a long-range strength (photon too massless particle). Building quantum theories gravity associated With big difficulties that's why gravitational strength often considered as a manifestation of the space-time metric. Let's say that within the general theories relativity gravity is equivalent to curvature space-time. More about these difficult things we let's talk later.

AT conclusion remains to tell, what at each elementary particles there is its antiparticle - his kind twin particle, possessing toy same mass, but charge opposite sign (if particle charge not It has, then her antipode bears opposite spin). When particles and antiparticles collide, their mutual destruction (annihilation) With highlighting huge quantities energy. More often Total final product annihilation are photons and pi mesons. O particles and antiparticles we too more not once we talk afterwards.

Echo Big explosion

And Tomlinson looked back and saw in the nightStars, tortured in hell, crimson rays.
And Tomlinson looked ahead and saw through the deliriumStars, tortured in hell milky white light.

Rudyard Kipling

AT end first chapters told about volume, what stars not distributed in space evenly, but form more or less compact structures (galaxies), which, in turn, are part of clusters and superclusters extending over tens of millions of light years. Our Galaxy (the Milky Way) is one such stellar islands and has about 200 billion stars (from 150 before 400 billion on different estimates). If a watch on the her With ribs, she is It has lenticular shape of a biconvex lens, and in plan, when viewed from above, it looks how flat disk co clot in center and outgoing from him spiral sleeves. The galaxy has a rather complex structure. It is customary to single out the core, or bulge (from English bulge - "convex, swelling"), disk and halo (galactic crown). Nucleusis a compact spherical component surrounding the galactic center, where there is a supermassive black hole with a mass of two to three million solar masses. The density of stellar population near the center of the Galaxy is very high: if in the vicinity sun on the 16 cubic parsec account for Total one star, then in center in one A cubic parsec contains about 10,000 stars. However, the density of stars in the bulge falls rapidly with distance from the center: at a distance of several thousand light-years, it is almost indistinguishable. The core is dominated by old stars with a low abundance heavy elements, a his weight evaluated in twenty billion solar wt.

More than half of the mass of the Galaxy (about 60 billion solar masses) falls on flat disk, inside whom sometimes allocate thin and thick part. Diameter galactic disk (and galaxies in overall) is 100 thousand light years, or thirty kiloparsec (30 kpc), and its thickness varies widely - from 300 to 3 thousand light years. AT areas center he thinner a to periphery noticeably expands. Disk contains a lot of young stars and dense clouds gas and dust - foci active star formations, which account for up to 10% of its mass. Galactic disk is wrong imagine as continuous homogeneous structure like a wheel or lenses, so how it breaks up into spiral arms, among which it is customary to distinguish two (sometimes four) big and lots of small. Sun located in 26 thousands light years (about eight kpc) from center galaxies and commits around him full turnover per 220 million years, flying through the void at a speed of 250 kilometers per second. If you count one revolution around the center in a galactic year, then the age of the solar system will be 20 galactic years - exactly so many turns she is had time wind up With moment his education.

Of course Sun not alone in his relentless circling - all stars disk revolve around the galactic center. The orbit of the Sun is almost circular and lies in plane of the galactic disk (only 20 light years away vertically), so study of the core of the Milky Way associated with significant difficulties. It is fenced off from us by disk stars that are closer to the core, as well as powerful gas-dust clouds, which not miss light from structures galactic center. Optical only the tail of the Galaxy is accessible to observations, and the most interesting is hidden from earthlings dense gas-dust veil. Here if would us somehow miraculously managed soar above plane of the Milky Way, we would see the mysterious bulge in all its splendor. To Unfortunately, such a prospect does not shine even for our distant descendants, because the Sun is in its orbital motion almost does not deviate from the plane of the galactic equator. AT our era flies in interval between spiral sleeves Perseus and Sagittarius, slowly approaching the sleeve Perseus.

Except flat disk and central swelling in areas core, Galaxy has a spherical halo that envelops the galactic lens like a cloud. Astronomers have long noticed that some stars do not swim measuredly and leisurely in a plane disk, a scurry about in most different directions, penetrating his through. Builds up impression, what they fill the whole spherical volume, where loaded galactic disk, forming a giant ellipsoid stretching for hundreds of thousands of light years. Halo inhabit old stars, which near 10 billion years from kind, then there is they twice older than the sun. One part of the stars prefers to live in splendid isolation, while the other is included in composition of the so-called globular clusters, of which there are about 200. In each of they contain from 10 thousand to 3 million stars, which is no more than 1% of all stars halo. Apart from ball clusters and solitary stars, in galactic crown discovered gas clouds and dwarf galaxies living at a distance of 150 pda from the Milky Way.

Although the total mass of the halo stars does not seem to exceed a billion solar masses, galactic crown much heavier our Galaxies. On the this is indicate some features of the rotation of the Milky Way and the nature of the movement of its satellites. Supposed, that most of the mass of the halo is associated with the so-called dark matter (or hidden weight). O problem hidden masses tells in chapter "And darkness came."

Our Galaxy is one of the spiral galaxies, which, according to the classification American astronomer Edwin Hubble, it is customary to designate letter S (from English the word spiral, which hardly needs translation). All spiral galaxies are made up of spherical and flat components, then there is from nuclei and disk, and disk It has expressed spiral structure. How rule major spiral sleeves happens two, but maybe count and more. AT dependencies from forms spiral branches and There are several subtypes of bulge sizes inside S-type galaxies: Sa, Sb, Sc, and Sd. AT In this row, the spiral branches become more and more ragged, and the size of the nucleus decreases. Spiral sleeves too may to be oriented differently: in some cases they begin directly from core, a in others cling to ends thick stellar jumpers, crossing central part galaxies. Such jumper called bar, and then galaxy hits in category SB (spiral + bar). galaxies With bar subdivided into the same four subtypes. There are serious reasons to believe that our Milky Way has a small bar, the extreme points of which are 3–4 pda from center, a on structure spiral branches and sizes bulge takes intermediate position between subtypes b and s.

Spiral galaxies are the most (over 50%), and among all the rest it is accepted identify elliptical, lenticular and irregular galaxies. elliptical galaxies nearly not contain interstellar gas and not have flat disk. By essence affairs, they are one continuous core, the shape of which varies widely - from an almost perfect sphere to an ellipsoid of varying degrees of oblateness. Hubble assigned them the letter E (elliptical in English), and expressed the degree of flattening in Arabic figures. So the way nebula E0 will be globular galaxy, a E6 acquires spindle shape. Lenticular galaxies are designated by the Latin letter L (from the English the words lenticular - "biconvex") and outwardly very similar to elliptical, since an impressive core prevails over a thin stellar disk, inside which, as a rule, no structural formations can be seen. Irregular galaxies - this is ragged ragged clouds, noticeably inferior in mass to other types of galaxies. More Total they similar on shapeless blots, inside which can sometimes discover unstable and short spiral arms. In classification Hubble they are designated how Ir or Irr (irregular - "wrong").

In addition to the variety of forms, many galaxies have a very noticeable activity. They explode and collide, drawing long jets of gas from their sisters' bodies and stellar substance, or, conversely, merge in a close embrace like germ cells under a microscope. Some of them radiate in the radio range and are thrown out of their active nuclei powerful jets length in several thousand light years. A textbook example is the radio galaxy Cygnus A. In optical rays, it represents yourself an object 17th stellar quantities in form two barely notable spots. But this is impression deceptively, because what in reality them luminosity in ten once more, how in our Galaxy. This system seems weak only because it is removed from us. 600 million light years away. However, despite such an impressive distance, the flow radio emission in meter range from swan BUT exclusively great and from time to time exceeds solar radio emission. But the distance from the Earth to the Sun is just eight light minutes...

The interaction of galaxies very often radically changes their structure. For example, two spiral galaxies may merge together, giving rise to elliptical a large galaxies, without grimacing, easily swallow small, thereby increasing your size. Our Galaxy is also far from a vegetarian. Astrophysicists believe that it formed in result mergers several relatively small galaxies, Yes and today milky Path keeps ear vostro, trying everyone in truth and by untruths attach eight dwarf galaxies in its immediate environment. BUT through 2–3 billion years to him destined fraternize With galaxy andromeda, which is at a distance of two and a half million light years and flies in our direction co speed of 120 kilometers in give me a sec.

About the Local Group, which includes our Milky Way along with the Andromeda galaxy, a galaxy in the Triangulum and four dozen smaller galaxies, we have already written. This gravitationally bound system, having a diameter of approximately 1 Mpc (megaparsec, million parsecs) is, in turn, part of a local supercluster in the constellation Virgo, which is 15 Mpc away from us. Meanwhile, only the core is located in Virgo local supercluster, but it itself, according to conservative estimates, stretches for 30 Mpc (on others data - on the 60), a his thickness is not less ten Mpc. Local supercluster It has form ellipsoid, a number galaxies, in German contained, approximately estimated at 20 thousand. In recent years, several dozen superclusters. Some of them are striking in their size, such as the giant a chain of galaxies stretching from the constellation Perseus to Pegasus and Pisces by almost 400 Mpc (more billion light years). it already not habitual ellipsoid, a quicker beads, strung on the branching a thread. AT hierarchy metagalactic structures similar conglomerates occupy an honorary first place.

What has been said not means what thesis Friedman about isotropy and homogeneity Universe turned out to be bankrupt. Despite on the strings galaxies, along and across interbedding Big Space, in volumes length in hundreds megaparsec space observable Universe all equals not It has dedicated directions. And only at decrease scale succeed make out cellular structures, where dense plots alternate With gigantic voids. Let's listen specialists:

...

The general structure resembles a honeycomb or soap suds, only it is more blurred, without a definite clear pattern. Cell nodes are formed by superclusters galaxies, and there are almost no galaxies inside the cells. The diameters of such cells reach several dozens megaparsec. trying introduce yourself structure Universe in these gigantic scale, important remember, what she is not static: Universe expands, her parts move away from each other, so the cells increase, as do individual superclusters galaxies.

Others words our world continuously evolves. Observations definitely testify what cellular structure all time deformed: "bridges" transferred between superclusters, lose weight and stretch, a walls cells little by little melt away and slowly spread. The universe is extremely non-stationary, it everything is growth and formation, and about this dynamics of it, discovered almost 100 years ago, it came time talk. But first - Few words about quasars.

This word is a transliteration of the English term quasar, which, in turn, represents yourself abbreviation term quasi-stellar radio source, what translated how
"star-like radio source". The first quasar was discovered in 1963 by the American radio astronomer Dutch origin Martin Schmidt. More precisely saying discovered he was three for years before and was listed in 3m Cambridge directory under number 3C 273 in the form of a faint star of the 13th magnitude in the constellation Virgo, and Schmidt is the first drew attention to the amazing features of its spectrum. Emission lines in the spectrum stars 3C 273 at first could not be identified with the lines of known chemical elements. In the end, Schmidt realized that this was not some new element at all, unknown to modern physics, but the lines of the most common chemical elements that so strongly displaced to red end spectrum, what have changed before complete unrecognizability. After a fair bit of brainstorming, Schmidt was able to identify hydrogen lines, ionized magnesium and some others elements.

But if the redshift is so large, then this means that the mysterious the object is moving away from us at a fantastic speed - more than 40 thousand kilometers per second. In this case, the distance to it must in no way be less than 620 Mpc, that is, almost 2 billion light years. (By red displacement define degree remoteness astronomical objects; this will be discussed below.) It does not look like the galaxy 3C 273 was, but to see a single star at such a distance, no matter how bright it shines, in basically impossible! After several more similar objects were discovered, shining brightly in the visible and radio range of electromagnetic waves, they were called quasars - star-like sources intense radio emission. AT our days known already over twenty thousand quasars, many from which brightly shine barely whether not on the all lengths electromagnetic waves - from x-ray to the radio band.

Another characteristic feature of quasars is the variability of their brightness with a period of several months what He speaks about emergency compactness these objects. If a would they were huge star islands like galaxies, their brilliance is by no means case could not change periodically, because to synchronize the "work" of billions of stars fundamentally impossible. Consequently, quasars - this is solid heavenly body, what, for example, are the stars. The synchronicity of change also indicates that they the diameter cannot be more than one light year. Looks very strange picture: the object is inferior in size to the galaxy by hundreds of thousands of times, and at the same time it shines like kind a hundred galaxies. And although them sizes, on all probability, noticeably outnumber diameter solar systems, on space standards this is all equals negligible few. By the way, in the radio range, no more than 1% of quasars radiate, and in the spectra of many of them, as already It was said that it is possible to detect not only X-rays, but also hard gamma rays. All quasars - very ancient education and located extremely long away, on the distances of hundreds of millions and even billions of light years, and the age of the most dilapidated quite comparable With age universe and reaches 13 billion years.

What same source so powerful electromagnetic radiation, and on the all wavelengths at once? Most experts agree that quasars represent are supermassive black holes that voraciously absorb matter from their surroundings. environment. Charged particle, captured gravity black holes, are accelerating before high speeds, which leads to intense electromagnetic radiation. Substance falls to the surface of the black hole in a narrowing spiral, forming an accretion disk, inside whom speed particle, overclocked field gravity, approaching to the speed of light, and the temperature in the central part of the disk reaches 100,000 degrees Celsius. Kelvin. By direction to periphery disk temperature falls, that's why quasar simultaneously radiates in broadest range electromagnetic waves - from infrared radiation and visible light to short-wavelength x-ray photons and tough gamma quanta. Powerful magnetic field captures charged particles and additionally twists them, forming jets - narrowly directed beams, a kind of fountains that shoot out from the poles at near-light speed and extend for hundreds thousand light years. Interacting With interstellar gas particles jets become source radio waves

In the era of quasars, the process of the birth of galaxies was in full swing, so the material there was plenty around. Supermassive black holes fed perfectly at that time, and therefore shone exclusively bright. However through some time them had to pull up straps and go on a diet. Thus, quasars can be considered as a certain stage in life supermassive black holes: not without reason them, how rule discover on the distances in thousands megaparsec, at most borders observable Universe. Not should forget, what light from most distant quasars flew to earthly observer many billions of years, so we see them as they were in their early youth. Necessary believe that today they have long since tempered their appetites and live peacefully in nuclei calm galaxies. But similar consideration It has and reverse force, that's why should take a closer look take a closer look to our nearest environment - after all The universe is known to be isotropic and homogeneous. You look, and there are nearby cooled down quasars-ghosts, sat down on starvation rations. Incidentally, such objects are indeed exist - remember about supermassive black holes in nuclei galaxies.

So that you, the reader, can imagine stock of vital young forces quasars, let's quote professors Moscow engineering physics Institute (MEPhI) FROM. G.Rubin.

...

By the way, energy, which average quasar radiates per give me a sec, enough would for providing the Earth with electricity for billions of years. And one record holder, with the number S 50014 + 81, emits light 60 thousand times more intense than our entire Milky Way with its hundreds billion stars!

Let's put an end to this major note and move on to discussing issues related to With the evolution of the universe.

Sir Isaac newton, formulated law world gravity, believed universe homogeneous, endless in space and unchanged in time (stationary). Space determinists represented yourself fabulous debugged and flawlessly functioning clockwork, where the uniform circling of the luminaries obeys strict mathematical laws. The stationary universe model seemed simple, logical, internally consistent, and therefore successfully survived to the beginning XX century. The space in which the course of the worlds took place was conceived as Euclidean, that is, flat. We will have a separate discussion about geometric knuckles in the following chapters, here I will remind you, the reader, what flat space is. In space Euclid, through a point lying outside a line, one and only one line can be drawn, parallel to the given one (the famous fifth postulate), and the sum of the angles of the triangle is 180 degrees. it most usual space, With which us account for collide daily. Regarding the age of the universe, there was no unity among the comrades: some believed world created in incomprehensible demiurgic act, a other thought what he exists forever. One word, enlightened public on the turn centuries lived in boundless stationary universe, existing unlimited for a long time.

However, infinity is scary. Reason yields to such categories, because they are not only devoid of visibility, but also sin with numerous inconsistencies. Of course, you can always mold a suitable metaphor, and then everything seems to fall into place. Was, let's say such beautiful eastern parable: "Far far away on the edge Sveta rises a huge diamond mountain, reaching its peak to the very sky. Once in a thousand years a small bird perches on the top of this mountain to sharpen its beak. When the bird is weaning mountain to the base, one moment of eternity will pass. Who argues, said gracefully and with taste, but in fact it is just an illusion of understanding. It is clear that sooner or later the bird will reach the base of the mountain, although it will have to spend a lot of time and effort. So the inconceivability of eternity has not gone away, it has simply moved into the unimaginable far

Parables are parables, but the model of the stationary Universe, infinite in time and space, there are much more serious shortcomings. If only things were limited psychological unacceptability of the category of the infinite, such a trifle can It would be nice to close your eyes. The trouble is that the postulate of a universe that exists unlimited for a long time, comes across on the unsolvable contradiction. Eternity can like a geometric straight line that extends in both directions - both in the past and in future. In other words, it has neither beginning nor end. But in this case, any arbitrarily chosen point in time (e.g. today) The universe already *exists* infinitely long. Consequently, all the processes taking place in it should long ago complete and the universe must remain in a state of some absolute equilibrium. However, astronomical observations irrefutably testify that the world is constantly evolving, and evolving rapidly. When we look through a telescope we look into the distant past of the universe and see that 10 billion years ago it was not the same as today. Please tell me, where does evolution come from if we have per back incalculable amount years? We already not talking about volume, what eternity on definition not maybe to be exhausted - on the then she is and eternity. Then how same she is managed crawl before our days?

The situation is no better with infinity in space. In 1823 the German astronomer Henry Olbers published work With criticism models endless stationary universe. He reasoned as follows. We first formulate three preconditions: 1) the extent of the Universe is infinite; 2) the number of stars is also infinite, and they are evenly distributed in space; 3) all stars have, on average, the same luminosity. Well what same, quite reasonable background. BUT now let's see, what at us succeed. Mentally placing solar system in center, Olbers divided all the space beyond it into a series of concentric layers, or spheres. The universe has become resemble an onion. Let layer B lie three times further than layer A. Then the volume of layer B will be 9 times greater than the volume of layer A ($Z^2= 9$), since the volumes of the layers increase proportionally the square of the distance of each layer from the center. If the stars are evenly "smeared" over all layers (premise 2), then layer AT, whose volume in 9 once more volume layer BUT, will be containin nine once more stars. FROM another hand, luminosity individual stars decreasing proportional to the square of the distance, from which it follows that the brightness of each star in layer B under the condition of their equal luminosity (premise 3) will be $(1/3)^2= 1/9$ of the brightness of an individual layer A stars. But there are exactly 9 times more stars in layer B! In other words, the luminosity of layers A and B will be completely identical, and the solar system will receive from these layers equal amount of light.

The same picture is true for all other layers, and since their number infinitely (premise 1), then the firmament must shine with an unbearable brilliance even at night. Sky will turn in one continuous gigantic Sun, what in reality not observed.

Olbers suggested what light, going to us from distant stars, weakens due to absorption in dust clouds located in its path. However, this counterargument is also untenable, since the clouds must gradually warm up and eventually begin to glow as brightly as the stars themselves. The only way to resolve the paradox Olbers (also called the photometric paradox) consists in the assumption that the number stars expressed as a final value.

Another paradox, received title gravitational paradox or paradox Seeliger, based on law world Newton's gravity.

Recall, reader, that, according to this law, bodies are attracted to each other with force, directly proportional work them masses and back proportional square distances between them. BUT because the stars not distributed strictly evenly on the fixed distances friend from friend, then swings density among stellar population will inevitably lead to the fact that sooner or later they will gather in a heap. Between by the way, this conclusion fair and for ultimate stationary Universe. Truth, myself newton thought what concept endless Universe allows to avoid this paradox because what endless number stars, distributed more or less evenly, never not pull together in point, So how in endless space No dedicated center. Preserved even his letter to Richard Bentley on the this topic.

Of course, Sir Isaac was mistaken, as his countryman Stephen Hawking wrote well in book "A Brief History of Time":

...

These reasoning - example Togo, how easily get into in a mess, leading conversations about infinity. In an infinite universe, any point can be considered the center, since according to both sides from her number stars endlessly. Only much later understood what more the correct approach is to take a finite system in which all the stars fall on each other, striving for the center, and see what changes will be if you add more and more stars, distributed approximately evenly outside considered areas. By law

Newton, additional stars, on average, will not affect the initial ones in any way, i.e. stars will fall at the same speed to the center of the selected area. How many stars would we no matter what, they will always tend to the center. Nowadays it is known that the endless static model Universe impossible if gravitational strength always remain forces mutual attraction.

So the way stationary model endless Universe turned out inoperable, because what not corresponded observant data. But if The universe has finite dimensions, the sacramental question immediately arises: what is located beyond its edge? The great German physicist Albert found a way out Einstein when in 1915 year published theory, which today called general theory relativity (OTO). He suggested what binder link between gravity and space-time is geometry. It was a real revolution in physics: in framework general theories relativity space-time thought not flat, as was considered from time immemorial, but curved under the influence of the masses and energies. This is easy to understand from a simple analogy. Material bodies bend spacetime, like to that how weighty ball causes deflection stretched films or rubber sheet. On the such twisted surfaces ball room two a smaller mass will no longer be able to move rectilinearly and uniformly: it will either roll into hole, educated heavy ball (will be attracted to him), or will change trajectory his movement. Similar way this is the case a business and With heavenly bodies: for example, the orbital motion of the Earth is not at all due to the gravitational attraction of the Sun, but by the features of the space-time metric. The shortest distance between two points in curved space will not be a straight line, but the so-called geodesic, more Total relevant straight lines in usual flat space Euclid. Thus, gravity in the general theory of relativity is considered as a consequence curvature spacetime, a matter not nested in empty box, where time and space live independently, but forms an inseparable unity with them. If from Universe take out all matter time and space too not will be.

Everyone has probably come across a geodesic line. When an airliner makes long flight (for example, from Moscow to Vladivostok), the dispatcher asks the pilots a route that does not run in a straight line, but along a great circle arc, which is justand will be a geodesic line. Thus, a way out of the logical impasse was found. Although the universe is finite, it is at the same time infinite, just as it has no borders surface spheres. Of course visually imagine this is not easy, but can resort to two-dimensional analogies. If a on the surfaces spheres live hypothetical flat creatures unaware of the third dimension, they will never discover the edges his universe, although she is It has quite final sizes. Surface spheres is described by the geometry of Bernhard Riemann, in which parallel lines intersect, and the sum of the angles of a triangle is greater than 180 degrees. The curvature of space depends on the average density of matter in the universe. At some critical value of density, the curvature becomes positive, and the space of the Universe closes on itself, forming four-dimensional hypersphere, whose analogue in three dimensions is the surface of the ball Or a baby balloon. The famous English physicist James Jean wrote: about this:

…

Universe, portrayed theory relativity Einstein similar bloated soap bubble. She is - not his interior, a film. Surface bubble is two-dimensional, and the bubble of the Universe has four dimensions: three spatial and one - temporary.

O geometry peace we we will talk more not once in subsequent chapters.

So, photometric paradox received beautiful permission. Universe Einstein finite (although not It has borders), that's why paradox Olbers removed myself yourself. However, despite on the breakthrough truly revolutionary character in understanding nature space and time his model remained stationary, that's why the gravitational paradox continued to hang over her like a sword of Damocles. Whatever was gravity in its essence - the interaction of gravitating bodies or the manifestation of a metric space-time, matter, filling finite volume, must inevitably pull to a point. To save his theory, Einstein was forced to introduce into the equations the so-called lambda term, the cosmological constant, which resisted the forces world gravity, effectively "pushing" matter. This enigmatic strength not generated any source, but was built in frozen in herself structure space-time. By Einstein universal strength repulsion in accuracy balances the attraction of all other matter. Need to say, that Einstein could not stand the lambda, knowing full well that it is nothing but a god from the car, ad hoc hypothesis (for this case), and subsequently called the introduction of the cosmological constant biggest mistake of my life. And indeed, very soon from her had to refuse. However, parting With nasty lambda passed quite painlessly.

Einstein's stationary model did not last long. Petrograd mathematician BUT. BUT. Friedman in 1922–1924 years earnestly showed what equations general theories relativity allow on extreme least several non-stationary solutions. Subsequently It revealed, what motionless static model Einstein inevitably becomes non-stationary, that is, the Universe must either expand or contract. In fairness, it should be noted that a few years before Friedman, in 1917, Dutch astronomer billem de Sitter too proposed dynamic model expanding universe, but he worked with ideal empty space, while Friedman twisted-spit real model, filled substance. About ideas Sitter (very fruitful and far ahead of his time) i I will tell a little bit later.

Friedman suggested that the world as a whole is not only homogeneous, but and an isotropic medium, that is, one in for which there are no designated directions. it was a very far-sighted thesis, because in reality this is the case way. Groups and clusters galaxies really create sensitive inhomogeneities, but only at relatively close distances. If we change at once scale and highlight in the volume of the observable part of the Universe (remember: it is commonly called Metagalaxy) a cube with a side of the order of 300 - 1000 Mpc (megaparsec), then we will see that large scale structure Universe is different high degree homogeneity and isotropy. Theory Friedman says what statics inevitably is replaced dynamics, moreover, the dynamics of a well-defined property - galaxies and clusters of galaxies do not have rights be in peace, but must scatter co speed, directly proportional distance between them. AT this is significant difference Friedman models from sitter script: in calculations dutch astronomer universe expands exponential, that is With acceleration.

Friedman's decision was first accepted with hostility (including by Einstein himself), but great physicist quickly revised your point of view. This is what we read in the article Alberta Einstein, published in 1923 year:

...

AT previous note I exposed criticism named above work (Work Friedmancalled "O curvature spaces." - *L. Sh)*. However my criticism, how I made sure from letters Fridman, reported to me dwarf Krutkov, based on the mistake in calculations. I consider Mr. Friedman's results to be correct and shed new light. It turns out that the field equations allow, along with static ones, also dynamic ones. (then there is variables relatively time) centrally symmetrical solutions for structures space.

A rare letter from which it is remarkably clear who xy is. The number one physicist was embarrassed to publicly admit his mistake, from which it follows that he did not consider his famous equations as the ultimate truth like the Old Testament decalogue (ten commandments, received Moses on the grief Sinai from hands in arms from creator Total being).

Friedman's solution meant that the Universe is not only finite in space, but also had a beginning in time. The beginning of the world must lie at a special point - a singularity (from Latin singularis - "special, separate"), where curvature space-time becomes infinite, and the very concepts of time and space lose all meaning. Matter squeezed into a point with zero dimension must have an infinitely large density and temperature. To wonder about what was before, what preceded singularity, not It has no sense, for no "before" quite simply not existed. The events that we are witnessing today have nothing to do with the fact that took place before the Big Bang, when the Universe suddenly fluttered out of non-existence. How successfully put it once upon a time famous domestic cosmologist I. B. Zeldovich, "It was time, when time not It was". That's why we we have complete right take advantage famous "razor Ockham" (not should multiply number entities in excess of necessary) in order to cut off inappropriate questions. Until the moment of "zero" (that is, the Greater explosion) not It was neither time neither space. Partly this is recalls pagan cosmogony of the ancients, when the motionless eternity is transformed into a lively historicaltime.

non-stationary solutions Friedman suggest three option development events. The first option: the curvature of space is zero (the average density of matter in the Universe in accuracy is equal to the critical density), that is, three-dimensional Euclidean space, an analogue whom - plane, expands unlimited. Second option: space It haspositive curvature (the average density of matter exceeds the critical density), that's why world represents yourself final on volume, but limitless hypersphere, inflating like a child's balloon or soap bubble. Because the the density of the substance is higher than the critical one, sooner or later the expansion will stop and be replacedcompression (expansion substances stop strength gravity). Third option: curvature space is negative (the average density of matter is less than the critical density), therefore, as in the first variant, the world expands indefinitely, only its shape is not flat, a represents yourself pseudosphere or hyperboloid, analogue which in two measurements is surface saddles. Such Universe described geometry Lobachevsky, where the sum of the angles of a triangle is less than 180 degrees, and through a point lying outside straight, can spend How many any straight, parallel given.

It is very curious that the theoretical calculations of Friedman and Sitter fell on the time when observational astronomy gradually accumulated evidence that our Universe, despite models Einstein by no means not stationary, a continuously evolves. All began With Togo, what American astronomer Weston Slifer on the throughout ten years (beginning With 1912 of the year) patiently photographed spectra extragalactic nebulae. At that time, no one knew that in reality they present yourself gigantic stellar islands like our galaxies and lie unimaginably far from the Milky Way. Slipher set out to calculate their ray speed, then there is install, approaching they to our Galaxy or, vice versa, are removed from her. AT their calculations he leaned on the a long time ago famous Effect

Doppler which the, I guess to you, reader, sign not So Good, how American astronomer. Therefore I'll do a little retreat.

Austrian physicist Christian Doppler opened Effect, named subsequently hisname, a very long time ago - back in 1842. Probably, it could be found earlier, but this is how a person is arranged - very often we look, but we do not see. Psychologists say what everything fault specifics our perception, which prefers push off from well-known things and frankly ignores everything unusual. Per trees man not sees the woods. How would there neither It was, but tell, what Claude Monet, one from founders impressionism, was first artist, who turned Attention on the famous London fog. Generations British even not suspected what in them British atmosphere, oversaturated with the smallest particles of coal, nothing happens quite special. But then a stranger appeared with an unclouded eye and immediately wrotepicture "Bridge waterloo (Effect fog)", which literally plowed haughty islanders.

FROM effect Doppler a business this is the case in accuracy So same. If a past you on highway a car rushes by with the siren on, then as it approaches, the signal tone sounds higher and higher, but as soon as she catches up with you, the sound immediately drops by a whole octave and then (on measure removal) becomes all more bass. That same most can observe on the station platform: the whistle of an approaching train stubbornly climbs up, but when it flies by, the tone of the horn jumps from high to low. The essence of the effect lies on the surface, for sound - this is alternation compressions and rarefaction air, a the distance from one region of compression to another is nothing but the wavelength. How the longer the wavelength, the lower the sound, and the shorter the wave, the higher the sound tone. If a source sound (in given case - train) moving on direction to to you, then on the unit length accounts for a greater number of waves - the wave "palisade" becomes more close. If the source is removed, then the picture is exactly the opposite. - length waves starts grow. So the way length waves, emitted source, depends not only from properties source, but also from his speed.

Light, like sound, also has a wave nature and is a vibration (or waves) of the electromagnetic field. Interval of frequencies perceived by the human eye (visible region spectrum), lies between red light With length waves 740 nm (nanometers, or billionths of a meter) and violet light with a wavelength of 400 nm. We perceive long-wave infrared radiation as heat propagating from heated bodies, and radio waves lying in the extreme right parts electromagnetic spectrum. Region short waves presented ultraviolet, x-ray and gamma radiation (as the wavelength decreases). Thus, both gamma rays, and visible light, and radio waves are in their physical nature by electromagnetic radiation and differ only in the length waves, or the frequency of oscillations per second. The higher the oscillation frequency, the shorter the length waves, and vice versa.

In the optical range, red light has the longest wavelength, followed by orange, yellow, green, blue, indigo and violet are the shortest wavelengths in visible areas spectrum. If a source Sveta moving on direction to us, then distance between crests next friend per friend waves will decrease a frequency fluctuations will increase accordingly. As a result, all lines will shift towards the purple end. spectrum by the same amount. We can say that the light of a star approaching us a little will turn blue. At removal object from observer arises oppositepicture: the interval between the crests of the waves increases, and the frequency of oscillations decreases. lines are shifted to the red part of the spectrum, and the light of the departing star becomes reddish shade. Thus, in the first case, we have a purple shift, and in the second - red. the value bias compare With position lines in spectrum motionless source.

Weston Slifer analyzed spectra 40 galaxies and came to conclusion what most of them are moving away from us, and at very high speeds - on the order of hundreds and even thousand kilometers in give me a sec. This fact his very intrigued because the where it would be more natural to detect a chaotic spread in the direction of their velocities. If you Flip a coin 40 times, it is highly unlikely that it will land heads up 35 times in a row. Such tricks are simply forbidden by the theory of probability. And the more dimensions spent Slifer, topics more strange took shape painting, for magnitude red bias increased from time to time. The situation was aggravated by the fact that the American astronomer, as we remember, had no idea about the extragalactic nature of his objects: he considered them nebulae, located in our Galaxy.

When in the mid-20s of the last century it was possible to prove that the Slifer nebulae in reality not what other how huge stellar islands, lying long away per beyond the Milky Way, breathing became easier. As soon as the object is found straightaway two unusual properties - abnormal speed and atypical location - can count, what between them exists some connection. work Slifera continued other astronomers, and through a short time at them in hands already was an impressive list of extragalactic nebulae with varying levels of red offset. First luck smiled in 1929 year our old acquaintance Edwin Hubble, who was actually a lawyer by education, and became interested in astronomy later. Comparing with each other the speed of galaxies discovered a simple pattern: than The farther a galaxy is located, the faster it moves away from us. Others words speed galaxies directly proportional them distance from earthly observer, which is expressed by the relationship $v = Hr$ where v is the removal velocity, r is the distance from the galaxy to the Earth, and H is the coefficient of proportionality, which subsequently received title constant Hubble on the first letter of his last name (Hubble).

I must say that Hubble was very lucky. He derived his law from observing galaxies that are only 1–2 million parsecs (megaparsec, or Mpc) from us, then as it is known today that at such comparatively small distances his law works, soft saying no matter, because the close galaxies "tied" forces gravity. Assuming what the most bright stars others galaxies (supernovae and new) have about the same luminosity, he compared them averaged absolute stellar value With visible glitter and in result received very big value coefficient - about 400-500 kilometers per second per megaparsec. In addition, at that time distances before nearest galaxies were calculated very not exactly: when in middle last century revised the scale of intergalactic distances, the nearest galaxies had to be moved twice as far, and the most distant ones increased their "separation" by 6–7 times. Is it any wonder then that Hubble was wrong in his calculations by almost an order of magnitude? The current value of its constant, calculated on the basis of modern methods and with help very sensitive equipment like orbital probe Wilkinson is 71 kilometers in give me a sec on the megaparsec.

Should have in mind what galaxies are moving chaotically, in most different directions, in volume including and across beam vision. Clear, what such own them velocities, called peculiar, should not be taken into account. Law Hubble works only With radial speeds, averaged on big number galaxies, located on the the same distance from us. Exactly on this reason he practically unsuitable for nearby galaxies, since their radial velocities are relatively small. Therefore, it is necessary to separate the velocity due to the Hubble offset, from individual (peculiar) ray speed, which maybe to be very significant. For example, local Group flies how single whole in side clusters Centauri at over 600 kilometers per second. But the farther away one or other galaxy, topics more her Hubble radial speed and topics less contribution in her value is introduced by the individual velocity of the galaxy. Thus, the most reliable law Hubble performed on the distances over 200 Mpc (200 million parsec), a for definitions distances before nearby galaxies better enjoy Cepheid scale.

It seemed would, the most accurate values distances law Hubble must give for most distant galaxies, but this is not at all so. A business in volume, what magnitude redbias at distant objects so significant what at calculations gives speed removal faster than the speed of light. Therefore, in calculating the velocities of the most distant objects (for example, quasars) need bring in amendments envisaged special theory relativity, and then formula acquires more difficult view (we her drive not we will become). Constant Hubble - fundamental constant, and importance her further refinement is obvious, since it is closely related to age our universe. If we mentally "scroll" the movement of galaxies back, we will come to to a moment when the distance between them was negligible. All matter will shrinkpoint, and the universe will cease to exist in its current form. As a matter of fact, Hubble's research, together with the work of Friedman, Sitter and other theorists, served starting point for creating the Big Bang model, according to which our world has there was a beginning in time. According to modern data, the age of the universe is estimated at 13.7 billion years.

Between by the way, from Hubble law stems curious consideration worldview character. Because the speed Sveta - maximum from all possible speeds, there must be objects that are as far away from us as that the light emitted by them will never reach the earthly observer. In other words, at astronomical observations at waves of any length, there is a certain physical limit beyond which penetrate in principle is impossible. The inexorable laws of nature outline the area accessible to our devices is an ideally empty, but insurmountable boundary, therefore it is completely meaningless to ask whether there are any objects or their there no. We them all equals never not we'll see for horizon events - very important concept in cosmology - cuts off native "our" from damn peace purebred where more reliable Soviet Iron Curtain. "There, under the clouds - eternity, "said the hero Saint-Exupery, flying at the helm of a dilapidated whatnot over a layer of continuous cloudiness, under which piled up rocky ribs Iberian mountains

Quantities red displacement, measured at distant galaxies and quasars, gave speeds so high that it was time to doubt the validity of Hubble's law. AT 1928 measured the radial velocity of the galaxy NGC 7619 and obtained a result of the order 3800 kilometers per second, and by the beginning of the 60s of the last century, objects were discovered that whose speed reached 40 thousand kilometers per second, that is, more than 1/8 of the speed Sveta. It is with this speed that the quasar ZS 273, discovered in 1960, is moving away from us. But this is were more flowers, because what already very soon, in 1965 found quasars With magnitude z = 3.5 (value z characterizes red bias spectral lines). it was monstrous, fantastic value, for red bias first quasars not exceeded 0.36 and was always less than one. The spectra of such quasars show distant ultraviolet line, moved out in visible part spectrum due to huge red offset. If a would not phenomenon red displacement, they would never not were discovered because the earthly atmosphere fully absorbs ultraviolet rays. Dutch radio astronomer Martin Schmidt, who worked in California and found this unique quasar, figured out what his speed is 81% speed Sveta (approximately 243 thousand kilometers per second). Over time, the number of such objects went for hundreds. The most distant quasar to date has been found at z = 6.43, from which it follows that the speed of its removal closely approaches the speed of light and equals 288 thousands kilometers in give me a sec. Distance before this quasar is 13 billion light years, the age of the universe at the time it emitted light was 880 million years (in our days - near fourteen billion years), a her the size in that time not exceeded 0.14 from modern. But what way gigantic objects, comparable on mass With our A galaxy that can move at such fantastic speeds? What strength gives them so incredible acceleration? To reply on the these questions, necessary figure out With physical the nature of the redshift.

After Togo how Edwin Hubble formulated mine law, from stationary models I had to give up once and for all. It became clear that the universe is a complex dynamic structure that is constantly evolving. The galaxies are moving apart cockroaches when you turn on the light in the kitchen in the middle of the night, and the rate of their removal increases proportional to the distance at which these galaxies are from us. If any a galaxy is twice as far away from us as another, then it will move twice faster. By the way, it should be borne in mind that it is not the stars that scatter, and not even individual galaxies, but clusters of galaxies. Let's say the galaxies that are part of the Local Group, not in a hurry to part with each other. Moreover, many of them, on the contrary, converge, as, for example, the Andromeda galaxy and our Milky Way, which fly on the opposite courses at a speed of 120 kilometers per second. The fact is that the expansion of the Universe as a whole does not affect (if we speak very strictly - practically does not affect) the movement objects connected by gravitational forces into a single system. The local group is just such gravitationally stable system.

But if the speed of the recession of distant galaxies is directly proportional to the distance to them, and a similar picture is depressingly monotonous, in which direction you look, there is a reasonable question: are we not in this case at the center of the universe? If solar system in this sense, frankly unlucky (as you know, it vegetates in the backyard Milky Way), then, perhaps, at least our Galaxy is the center of the universe? Such a conclusion would certainly warm the soul of many, because anthropocentrism sits in our livers. Alas, have to you, reader, disappoint: first peculiarity global expansion of the Universe lies precisely in the fact that it does not have a dedicated center. Friedman understood this when he offered his model to the most respectable public. He proceeded from two obvious parcels: firstly, Universe isotropic and homogeneous on the large distances, and secondly, the same statement is true for any other her points. In other words, in whichever of the galaxies the observer finds himself, he will see everywhere amazing picture of the expanding universe, and his own galaxy will appear to him motionless center peace.

This is easy to explain with an example. If you take a rubber cord tied to it with knots and stretch it, suppose twice or three times, then the distance between the pair neighboring nodes will increase exactly the same number of times. If we select one node in quality points reference, then speed removal others nodes will be grow directly proportional to their distance. You can also refer to the two-dimensional model. Let's take a children's balloon and put marks on its surface. As the balloon inflates the marks will begin to spread in different directions, but at the same time none of them will occupy privileged central provisions, a distances between them start grow according to the same proportional law. So, the first feature of the extension lies in the fact that all its subjects (that is, galaxies) are completely equal, and dedicated center, from whom they scatter, absent.

The second feature of the extension is already familiar to us. Not only the galaxies themselves (not to mention already about individual stars or planets), but even them clusters present yourself stable systems tied by gravitational forces, so the expansion of the Universe does not affects. When stretching the rubber cord, the distances between the knots increase, but not at all because they slide along the thread. It's all about elastic properties rubber, a themselves nodes run away nowhere not think.

From here follows and third peculiarity extensions Universe. His often represent as a recession of galaxies in space, which is completely wrong, because in given case missing traffic "something in something." Can to tell, what this is swelling space itself, although such a statement would be only a metaphor, because space Universe does not expand in some external on towards him volume. To use the terminology of Immanuel Kant, this is an extension of the space an sich, then there is in yourself himself. Imagine visually similar impossible, for for this had to would draw closed on the myself sphere in fourth spatial measurement.

So the way from epochal discoveries Hubble and works theoretical physicists it followed that our universe, in all likelihood, has a finite volume and was born in some zero-point of time. Or, to put it more strictly, at the point "zero" happened birth triplets, for matter, space and time not may exist apart. It remains to figure out exactly how events developed at this special singular point. For the first time, the Belgian astronomer Georges Edouard Lemaitre was seriously concerned about this issue, who in 1927 suggested that at zero point in time matter and energy future Universe represented yourself some superdense clot - his kind

"cosmic egg". AT strength unknown reasons happened catastrophic explosion, scattered matter in all hand, and fragments this world cataclysm we are still observed in the form of a recession of galaxies. Lemaitre's model of the universe was physical analogy theoretical calculations Friedman or Sitter, but at this turned out to be simpler and more understandable than the abstract constructions of highbrow mathematicians. That's why English astrophysicist Arthur Stanley Eddington became its zealous propagandist, and after some time, it was willingly adopted and thoroughly developed by the American scientist Russian origin George Antonovich Gamov. FROM his light arms non-stationary model of the hot universe was called the Big Bang theory and after the inevitable but necessary retouching, it remains in great use to this day. Gamow proposed his script in 1948 together with colleagues Alfer and Bethe, which speaks of Georgy Antonovich's good sense of humor, since the names Al-fer, Bethe and Gamow marvelous remind first letters Greek alphabet. Sometimes theory Gamow called a, I, y-theory on the what, apparently he and counted.

Judging on calculations Gamow, temperature and density inside space eggs must were surpass all conceivable limits, but already through one minute after The temperature of the Big Bang dropped to 109–1010 degrees Kelvin, and protons and neutrons, remaining after annihilation With antiprotons and antineutrons (about this more will be discussed below), began to combine into nuclei of deuterium, tritium, helium and lithium. This process received title primary nucleosynthesis, and Gamow managed show, what the ratio of hydrogen and helium observed today (approximately 75 and 25% respectively) arose in the first seconds after the Big Bang. According to his calculations, the stars for all time the existence of the Universe could not "produce" more than 1% of helium, which is not at all like those 24–25%, about which unambiguously they say astronomical observations. So Thus, the theory of the hot Universe received one more additional argument in its benefit.

All this is very good and even wonderful, but the time has come to take the villains to the nail and tough to ask in the spirit of Mikhail Zhvanetsky: and why, exactly? Why did not know grief and sadness, the cosmic egg suddenly became unstable and exploded? Is it really such a sensitive ephemeris that crumbles to dust at the slightest touch? If a the egg was still a stable structure that lived comfortably for many billions of years, then it should be clearly explained what unknown forces prompted the poor thing to do a series of sudden metamorphosis.

Questions, needless to say, are extremely difficult, so theoretical physicists proposed in their time quite a few models, in which not washing, So skating tried flatten ends With ends. Here, for example, is the so-called hyperbolic scenario: the Universe was originally represented yourself cloud extremely sparse gas, which the gradually condensed and warmed up under influence gravity forces. When gas pulled together in dense clot, centrifugal action high temperature and pressure broke gravitational contraction and the substance of the young universe splashed in all directions, like how a jet of hot steam flies out from under the lapped lids kettle on fire. Thus, the Universe begins its life in almost absolute vacuum, and then, stepping over phase maximum density, again returns in condition emptiness. hyperbolic Universe described geometry Riemann, a her radius curvature fluctuates over a wide range - from a minimum in the period of compression to a maximum in the period extensions. It begins with emptiness and ends with emptiness, and the stage of the cosmic egg turns out short intermediate stage on the background irreversible polar change. minus such models turn out irreversible states, spaced apart on different ends timeline.

Hypothesis pulsating Universe deprived these shortcomings. She is practically matches co second decision equations Friedman (cm. above) and represents yourself eternal oscillatory process between state ultrahigh density and phase maximum expansion. When the forces of universal gravitation (provided that the average density matter above critical density) stop expansion galaxies, red the displacement will change to purple and the galaxies will again rush to each other in their arms. Chemical reactions will also change their sign, and heavy elements will begin to decay into more simple. In other words, when the universe shrinks into a point again, it will again be consist from one hydrogen.

Based on modern ideas, the Universe after its birth from singularity experienced a short-term stage of ultra-fast inflation - the so-called inflationary period (discussed in the next chapter). After the end of inflation she is passed in mode proportional Hubble extensions, which transition and perceived by us as the Big Bang. At the turn of these two epochs mysterious field with negative pressure, driving no less mysterious inflation, ordered a long to live, and the released energy gave rise to a boiling broth of elementary particles, which warmed up newborn universe before beyond temperatures.

However, models are models, but still I would like something more real, which can be felt by hand. Redshift, no doubt, makes you think a lot, but it's just geometry, and not very easy to understand. But if it were possible to find some material trace of the hot beginning of the Universe, then it would be completely different talk. G. A. Gamov, the author of the Big Bang theory, back in the late 40s of the last century predicted what Universe must to be evenly filled radio emission millimeter range with a temperature of 25 to 5 degrees Kelvin. The matter remained small - discover such radiation.

AT 1964 year american physics Arno Penzias and Robert wilson, employees laboratories Bella, experienced most sensitive on the that moment detector microwave waves (microwave detector). To be fair, it should be said that they they were not looking for some unknown radio emission, but were engaged in debugging equipment for work on program satellite connections. For testing was selected wave length 7.35 centimeter, which was not emitted by any of the known sources. Antenna included in disposal Penzias and wilson, was wonderful and that's why they were extremely surprised when discovered what she is constantly fixes outsider radio noise, from which could not be got rid of. This noise was monotonous and even and did not depend on neither from directions antennas, neither from time days, Consequently, his source must located outside the earth's atmosphere. Moreover, it did not change even during of the year (a after all Earth flies on orbit around sun), from what should to conclude, what source radiation located not only per outside solar systems, but and per outside the Galaxy, because as the Earth moves, the detector changes orientation in space. Ironically, two other Americans, Robert Dicke and Jim Peebles, prepared search background isotropic radiation With temperature below ten degrees

Kelvin quite purposefully, but Penzias and Wilson, quickly realizing what was happening, reported About our results before.

Stephen hawking writes on this about:

...

Dicke and Peebles prepared to search such radiation, when Penzias and wilson, knowing about the work of Dicke and Peebles, realized that they had already found it. For this experiment, Penzias and Wilson were awarded the 1978 Nobel Prize (which was not entirely fair, if remember dick and Peebles, not speaking already about Gamow!).

Subsequently microwave background radiation managed register and on the others lengths waves - from 0.5 millimeter before several dozens centimeters. Outcome long-term observations was reduced to the fact that it has a thermal nature and corresponds to radiation absolutely black body at temperature 2.7 degrees Kelvin (exactcontemporary meaning - 2.725 TO). His spectrum not similar on the spectrum radiation stars, radio galaxies and other possible sources, and its intensity is almost identical when observing different parts of the celestial sphere, that is, it is isotropic and homogeneous, which is required prove. Soviet astrophysicist AND. FROM. Shklovsky proposed name mysterious radiation "relic", and since then the term has been widely used, although official his name - space microwave background.

What is relic radiation, and where did it come from? When about 14 billion years back in result monstrous explosion were born space, time and matter, Universe at first was boiling soup from protons, electrons, photons (light quants) and neutrinos, which violently interacted between themselves. All space newborn Universe It was filled solid opaque environment in form high temperature ionized plasma. As the universe expands, the temperature fell, and when it dropped to 3000 degrees Kelvin, the formation of stable atoms. There was, as astrophysicists say, the separation of radiation from matter, because it practically does not interact with neutral atoms. The universe has become transparent to radiation, and it was able to propagate freely. Sometimes this moment is called the epoch of the last scattering. The radiation temperature continued go down in progress further extensions universe, but his spectrum preserved without changes to the present day as a reminder of the hot days of our world. Here are the remnants former luxury and discovered future Nobel laureates.

Not will be exaggeration to tell, what opening microwave background had fundamental meaning and on his importance quite comparable With discovery expansion of the universe. The last nail was hammered into the cover of the stationary model. In second half XX century hot model Big explosion turned in solid full theory. Academician I. B. Zeldovich So said about this in 1984 year:

...

Theory Big explosion in real moment not It has any notable shortcomings. I would even say that it is as securely established and true as it is true that the earth revolves around the sun. Both theories were central to the picture. his universe time and both had a lot of opponents who claimed what's new ideas, mortgaged in them, absurd and contradict sound meaning. But similar speeches not in able hinder success new theories.

Of course, the respected academician was a little cunning, because even on the Sun there are spots, and theory Big explosion in this sense by no means not exception. Highly soon

It revealed, what, despite on the all my predictive force, she is too not deprived shortcomings, but about this - in next chapter.

Comprehensive inflation

In needle-shaped plague glasses
We drink the delusion of
reasons We touch small hooks,
How easy death, quantities,
And where the spillikins clashed,
The child keeps silence - Big
Universe in cradle
At small eternity sleeping.

Osip Mandelstam

Literally translated from Latin, the word "inflation" means "swelling". hardly needed explain, what overproduction paper of money or other payment funds, allowing endless replication by means of a printing press, leads directly to the aforementioned swelling for empty paper, standing pennies, immediately comes in contradiction with the actual supply of goods. However, the citizens of our country are familiar With inflation not hearsay: With most start 1990s she is hanging above head everyone law-abiding Russian like Damocles sword, a monthly summaries cheerfully report as far as lose weight his wallet per reporting period.

Astrophysicists economic turmoil occupy few, but contemporary cosmology With willingness took on the armament solid term, along the way returning to him original meaning. If in economics inflation is just a beautiful metaphor, then in cosmology, it is understood as a real physical process - a rapid inflation resurfaced from singularities newborn space. it regular and a necessary stage in the history of the very early universe, fundamentally different from who replaced his trivial extensions, about which detail told in the previous chapter. The question immediately arises: why did physicists need to introduce additional entity, if old kind theory Big explosion, seemed would, well explained all the observed facts? After all, even the famous English scientist Fred hoyle, heretic from astrophysicists and original thinker, diligently developing theory of a stationary universe, eventually gave up and accepted the concept of the Big explosion.

The fact is that within the framework of the traditional model, several solutions could not be found. very important cosmological problems. Before Total this is So called problem horizon particles and problem flatness. Except Togo, standard model not gave response on the question, what It was before Big explosion, and not was able explain sizes observable universe (if the Big Bang theory is correct, then the universe should be much smaller). These annoying inconsistencies, like splinters, stuck out of the body of a standard theories, and many cosmologists openly turned a blind eye to them, believing that with the passage In time, they will sort themselves out on their own. However, events turned in such a way that insignificant little things increased fundamentally different scenario origin our peace. Something similar in his time happened With outstanding German physicist Max Planck, who was tried to be dissuaded from theoretical physics because this science is almost complete. Only individual specks darken its bright horizons, the teacher taught him by life said to him, why waste your best years on stupid glossing? Planck, as you know, did not listen: he soon proposed quantum hypothesis and deduced his famous constant, thereby laying the foundation for a new one, non-classical physics.

Let's analyze the inconsistencies of the Big Bang theory in order. Let's start with the horizon problem particles. Astronomical observations show what Universe exclusively homogeneous in big scales. Temperature relic radiation, how we remember averages about 3 degrees Kelvin (2.725 K), with temperature deviations from middle values on various directions absolutely insignificant - they not exceed one hundred thousandth (10^{-5}). distances, available modern telescopes, fit into a value of the order of 10 billion light years, and in these spaces we we observe exactly the same thing - a striking "smoothness" of density contrasts. According to modern concepts, the true size of the universe is many times greater than it the observable part, which is usually called the Metagalaxy. Since the beginning of the world took place about 13–14 billion years to that back, light from distant objects elementary not managed before us get there - to him simply not enough time. Stars and galaxy, located per horizon events (if such there available), fundamentally inaccessible, because the speed of light is the maximum possible of all speeds. But inside horizon all particles causally connected friend With friend, So how they a long time ago already managed exchange between yourself necessary information.

The catch is that the Big Bang theory fails to explain how this exchange could take place. The horizon grows (and has always grown) at the speed of light, and interaction between particles in complete compliance With theory relativity inevitably carried out at slightly lower speeds. Cosmologists write: horizon particles always will be expand faster mutual distances between two trial particles. It turns out, what thermal equilibrium (a his Existence - indisputable fact) could in no way be achieved within the framework of the standard modelper expired 14 billion years.

When the universe was 300 thousand years old, the temperature of the plasma dropped significantly, and began education neutral hydrogen. Radiation separated from substances and photons were able to propagate freely in all directions. This the moment of time is usually called the epoch of recombination, or the epoch of the last scattering. It is clear that the size of the horizon at that distant time was much smaller than the current 10 billion light years and was approximately one megaparsec (one Mpc). So Thus, at the time of recombination, the thermal equilibrium could set on a scale not exceeding 1 Mpc. Today plot has such a size in the firmament angular the size near 2 degrees, Consequently, we entitled expect notable hesitation temperature of the relic radiation filling the Universe. However, astronomical observations show high degree isotropy on the all corner scales: temperature differential, how we remember not exceeds three hundred-thousandths (3×10^{-5}).

Apart from Total other things in framework standard cosmological models remains the mechanism of the initial push is incomprehensible. What force set the worlds in motion? Perhaps the universe arose as a result of the monstrous power of thermonuclear explosion of unknown nature? After all, the standard cosmological model which was created by the works of G. A. Gamow and other scientists, and is called the theory of the Big explosion. But at closer examination immediately clear: explosive mechanisms give practically nothing. In an explosion (chemical or thermonuclear - no value) It has) arise difference pressure and heterogeneous distribution substances: in one more flies to its side, less to the other. In addition, there must be a special dot - explosion center.

AT real same Universe nothing similar not observed: she is on the rarity is homogeneous, and some distinguished point, which could be identified with the center, is not is found. Already mentioned FROM. G. Ruby, Professor MEPhI, writes on this about:

...

It's the same as if our Earth had an ideal shape of a ball with "mountains" not over 40 meters high. For comparison: the diameter of the Earth is approximately 1.2 x 107 meters. Difficult It was would then believe in her accident origin.

Not less hassle at standard cosmological models arises and With So called the flatness problem. This somewhat clumsy turn means that we we live in an almost flat world, described by the geometry of Euclid, which everyone studied in school. How known physical space maybe to be twisted under influence gravity. Strictly speaking, Einstein's general theory of relativity considers gravity as a kind of reflection of the space-time metric. Imagine visually twisted three-dimensional space not easy, but this is can without labor do, referring to the corresponding two-dimensional analogs. The surface of a sphere represents yourself closed two-dimensional space ultimate area, which, topics not less, notIt has borders. Hypothetical inhabitants such peace (this is flat creatures, third dimension unknown to them) can move in any chosen direction, time after time crossing alone and those same points, but nowhere not discover the edges his Universe. Sphere With growing radius will be not bad analogue expanding closed three-dimensional space. Such a non-Euclidean surface is described by Riemann geometry, and the sum corners triangle on the her more 180 degrees. non-Euclidean geometry Lobachevsky is realized on the surface of a hyperboloid or pseudosphere - a complex curved structure, reminiscent surface saddles. Such universes will open, a sum corners triangle in them will be less than 180 degrees. Finally, available intermediate option
– non-curved plane described by Euclid's geometry. As in the case of complex surface of Lobachevsky, this flat world will be open and infinite in area. Similarly, our three-dimensional space, in which we we live.

The space of the real Universe at large distances comparable to the horizon particles, as already mentioned, is almost flat. Of course, this does not exclude areas local curvature, especially near large gravitating masses, but in cosmological On a scale, the deviation of the geometry of our world from the geometry of Euclid is absolutely negligible. Geometry space most direct way tied With size, denoted Greek letter ?, which is attitude middle density matter of our world to a critical density. If a ? is equal to one, then our Universe is perfect flat structure. If a Y more units (density our peace above critical), then Universe on achieving some maximum radius will start shrink under action gravity. AT this case early or late Big explosion will be replaced by the Big Crash (or Big Crunch), and the Universe will again turn into a point and will disappear in singularities. If a ? less units (density Universe below critical), the world will expand indefinitely, and the density of matter will become gradually fall.

Measurements carried out in recent years have shown that this value is very close to unit, although, most likely, it is not exactly equal to it (measurements are not yet completely reliable). This is where the notorious problem of flatness comes into play. Knowing approximate parameter value ?, can without big labor calculate, what must be the initial conditions of the very early universe to lead to today's observed values. And immediately shaped miracles are revealed. Let's quote M. AT. Sazhina, author fascinating books "Modern cosmology in popular presentation":

...

Let's take the parameter approximately equal to one, say 0.5 or 1.5. Let's see now how it should be in different epochs of the evolution of the Universe that were before our era. AT era of recombination difference Q from units already not must exceed 0.001. A greater difference would lead to what is today? would be equal to 10 or, say, 0.1, which easily measurable. AT era nucleosynthesis difference Y from units not must exceed 0.00000000000000001. AT more early era quark-gluon plasma difference Q from unity
"hidden" in 21 decimal places. At the Planck moment (this is the very beginning of our world, which we'll talk about later. – *L. Sh.*) this difference was expressed as a value of 10^{-60}. Where may take such initial terms?

Others words develops impression, what initial options were fitted with unprecedented precision: otherwise, we could not have managed for any price would get today's quantities indicator ?. Not by chance some astrophysicists they say about thin at the construction site parameter density. What and talk, painting unpleasant, making you seriously think about the creator of all things. Meanwhile, rigorous science somehow does not it is fitting to engage in empty arguments about a higher mind. This is the lot of philosophers and theologians. But whether there is a possibility not give away cosmology on ransom theologians?

I'm in a hurry you, reader, calm down - such possibility us in complete measure gives inflationary scenario birth universe, about which already for a long time it's time talk more. He easily and at ease removes and problem horizon, and problem flatness, and a bunch of other problems, under the weight of which the classical model was exhaustedBig explosion.

So, what is cosmological inflation and how does it differ from standard inflation? extensions, which we continue observe today in form red bias in spectra of distant galaxies? Inflation is a period of catastrophically rapid inflation space in the initial phase of the life of our Universe. Say it was a bloat swift and fleeting - to say nothing. Its duration is within vanishingly small terms: inflation started when age Universe was 10^{-43} seconds, and ended when it reached 10^{-37} seconds. At the beginning of inflation The universe was a little more than 10^{-33} cm, which is comparable to the Planck length, and at the time of its graduation was equal to about 0.1 cm (in others inflationary scenarios this magnitude ranges from one to thirty centimeters), that is, its diameter has grown by at least 10^{27} times.

Easily see, what initial extension young Universe happened co speed, repeatedly exceeding speed Sveta, because the Planckian length and time are interconnected: in 10^{-43} seconds, light has time to travel a distance no more, how 10^{-33} cm. Really we finally refuted most Einstein? Not we will hurry, reader. In reality, no contradictions here no and in remember, for the theory relativity limits the speed of light only the movement of material bodies, but says absolutely nothing about the rate of expansion of the space itself as such. Bye particles substances continue move co speeds smaller how speed light, the space surrounding them is allowed to swell arbitrarily fast: the speed its inflation is limited only by the amount of available energy, providing mentioned inflation.

Between by the way, introduction the one and only additional parameter - exponential inflationary expansion - automatically solves the damn problem horizon. AT his time we postulated what horizon always growing faster, how increases distance between two dots (or two particles) in space. However soon after birth Universe this is condition, obviously, not was performed. Imagine a tiny young Universe of the order of the Planck length - a little bit more 10^{-33} cm. Inside this domain more *before start* inflation managed settle down thermodynamic equilibrium and causal connection. When comes phase inflation,

space is rapidly accelerating, literally swelling by leaps and bounds, as a result of which a microscopic homogeneous area almost instantly monstrously grows in size. The domain volume grows much faster than the horizon distance. By the end inflation he is about one cm3, and inside this areas Universe is
"smooth" without notable contrasts of density, temperature and pressure. Further inflation gives way to standard expansion, and the particle horizon continues its leisurely grow, reaching to our time quantities order 1028cm. At this all particle, filling the observable part of the Universe, even before the start of inflation, they managed to establish between yourself causal connection. domain, overgrown in progress standard extensions, saves then condition, which formed in time inflation. Cosmologists they say,what all contemporary Universe located inside one causal areas.

Similarly solved and problem flatness. Today space our The universe is practically flat, but before the epoch of inflation, the parameter ? could be significantly different from unity in any direction. Whatever the curvature of the world near the point "zero" in in the end, we still get an almost flat model, because inflationary swelling smoothes density contrasts. This is easy to see with a simple example. Suppose that the density parameter before the start of inflation was noticeably greater than unity (? › 1). Then we we obtain the topology of a closed space, that is, the Universe is equivalent to a surface spheres. At inflation ball his radius growing, and if choose on the his surfaces small enough area, its curvature will be practically indistinguishable from zero. In the end ends surface Earth seems us absolutely flat. If a same recall,that in some models of inflation (we will talk about various inflationary scenarios a little below) initial tiny domain, comparable With Planckian length, bloated before astronomical quantities 101000cm, then observable Universe (or Metagalaxy), diameter which about equals 1028cm, will be make up insignificant part of the giant Megaverse. It is clear that in this case the microscopic area, not exceeding Yu-1000pieces gigantic ball, will be perceived how perfect flat. So the way No no need postulate special initial conditions that subsequently ensured an almost zero curvature of the universe. Parameter density could take any values around the point "zero", So how comprehensive inflation inevitably smoothes all bumps and will do space practicallyflat.

Let's get back to beginning inflation, in era very early universe, when her age was 10-43 seconds. What forces dispersed space to unimaginable speeds and increased his volume on the orders orders? To reply on the this tricky question, scientists had to introduce an additional concept of the inflaton field, which is often also called the scalar Higgs field and the fake or false vacuum state. You should not be afraid of this, because in order to explain the mystery of the hidden mass and dark energy(talking about these phenomena ahead of us) anyway, one way or another, you will have to resort to new fields unknown modern science. In the wild high physics we not climb, because the adequately figure out in these things without very complex mathematical apparatus not seems possible. Note only, what the hypothetical inflaton field has very strange and even slightly frightening characteristics.

Let's turn to visual example, so that in alive images illustrate status. Imagine a snow-covered mountainside full of bumps and local elevation changes. You roll up the snowball and send it down the slope. If a snow enough wet, snowball will start fast increase in sizes, bye not turns into a huge lump. The process develops exponentially - the larger the diameter snowball, the faster it grows. Our hypothetical slope ends in an abyss, and whenthe snowball reaches the edge of the cliff, then in full accordance with the laws of physics it will fly vertically way down With growing speed. Caught on the day, he to smithereens will break and part kinetic energy snowy coma will leave on the heat environmental environment.

Now we'll be back to inflaton field With his mysterious characteristics. Firstly, this is a scalar field, that is, a field that is not oriented in space in any way, in difference, let's say from electromagnetic. AT German missing power line, a his tension everywhere is the same. FROM some reservations his can liken homogeneous substance like viscous spreading honey. Second, the inflaton field characterized extremely strong negative pressure, which literally "pushing" substance, overcoming strength gravity. AT standard hot models At the Big Bang, the density of matter falls as the size of the universe increases, which quite naturally, So how energy density determined cash energy, divided by volume. But the inflaton field (that is, a fake vacuum) behaves paradoxically: his energy density on measure inflation remains permanent, so the energy manager swelling space, not only not decreases a on the contrary, it grows exponentially. However, nothing lasts forever under the moon - a state of matter with growing negative pressure is extremely unstable, and therefore must inevitably change mode extensions. Phase inflation swiftly coming off on the No, and all potential energy of a fake vacuum turns into a boiling soup of newborns elementary particle, warmed up before highest temperatures. Others words With ending era inflation is born ordinary matter in form hot plasma.

Let's take another walk on the snow-covered mountain slope and play game again. snowballs. AT this comfortable models analogue inflaton fields, filling all space, there will be snow on the slope. Thanks to random quantum fluctuations, our the field can take on a variety of values in different areas. Snowball formation is exactly such quantum fluctuation. Bye snowball rests, nothing nothing remarkable happens, but as soon as he moves down the slope, he immediately starts swiftly grow. Inflatonic field, inflating newborn fluctuation, tends to take a position in which its energy is minimal. Exactly that same most going on and co snowy lumpy: losing energy and monstrously swollen he reaches finally the edges cliff and falls down in abyss, a all accumulated them energy is being transformed in kinetic energy scattered particles. Bye snow com travels up the mountainside, inflation keeps picking up all the time, but costs to him touch bottom gorges, how energy inflaton fields shrinks before minimum for fall more nowhere. going on warming up universe, and how once this moment perceived by us as Big Bang.

The plateau on which our snowball rolls is by no means a smooth polished table without bitch and hitch, a surface, having where more difficult relief. Local elevation changes in the form of various kinds of bumps and unexpected obstacles inevitably perceptible disturbances in the snowball trajectory. In addition, such lumps (read - quantum fluctuations) there are a great many on the slope: some lie closer to the cliff, others are further away from it. And if individual snowballs succeed relatively freely slide down straight ahead way down, then other doomed dodge and jump "onvalleys and hills", getting stuck in pits and deep potholes for a long time. They lead exactly the same myself and real quantum fluctuations - embryos future universes: alone from them experience short-term inflation (inflation, as we remember, continues until since, bye snow com moving on plateau), other bloated before now since, a third instantly collapse, not having time how should grow up. So the way in our a whole ensemble of universes is at your disposal instead of a single one, each co their set unique properties.

This scenario, received title eternal, or chaotic, inflation, was proposed in the mid-80s of the last century by an outstanding American astrophysicist Andrei Linde, our former compatriot. Among other things, the model of eternal inflation wonderful topics what allows get rid of from curses contemporary cosmology - the anthropic principle. However, we will talk about the anthropic principle in next chapters, here same note only, what fundamental constants (gravitational constant, electron mass, etc.) and the very laws of nature that govern behavior our peace, amazing way allow occurrence complex structures generally and reasonable life in particular. If a them value slightly tweak (at all a little bit, on the insignificant share percent), Universe will be transformed radically. Let's say at otherwise ratio masses proton and electron education any complex structures will become fundamentally impossible. Between topics observable ratio - naked empirical fact, not derivable from theoretical constructions. As if someone wise, far-sighted and prudent, having carefully weighed all pro et contra, specially selected the values of the fundamental constants in such a way that unfriendly space became "hospitable" for person. BUT here idea about countless multitude universes, divergent on their parameters, automatically removes this problem.

In fairness, we note that the hypothesis of an inflationary stage in the history of the early Universe was first expressed domestic scientists E. B. Gleaner and BUT. BUT. Starobinsky more in 60 - 70s years of the past century, but stayed to unfortunately unclaimed by the scientific community. The term "inflation" was coined by the American physicist Alan Guth in 1981, and he also built the first inflationary model based on a kind of phase transition that caused the supercooling of the young Universe. Not here place to analyze the Gutian scenario in detail, since it quickly became clear that he does not work, as it gives a very inhomogeneous Universe in the final, which in reality not visible. But the model of A. D. Linde was devoid of these shortcomings, than immediately has gained unprecedented popularity: if before the inflationary scenario is very often accepted with hostility, today most physicists and astronomers have joined the ranks of his supporters. From beautiful, but shaky hypotheses inflationary Start Universe turned in full-blooded scientific theory, allowing experienced check. Cosmology, former before recent time discipline in significant degree speculative little by little becomes strict experimental science.

How we remember theory inflation postulates Availability insignificant changes in density of matter in the early universe. Since the volume of the newborn world is comparable to dimensions elementary particle, reasonable suppose what quantum fluctuations were playing at that time a very significant role. Werner Heisenberg's uncertainty principle says that we cannot simultaneously calculate the exact position of a particle and its momentum (product of velocity and mass). In other words, the energy and position of the particle never not may to be measured exactly and this principle in complete measure apply to first moments life Universe (ball winding on slope, a not rolling straight ahead way down). Total Effect quantum fluctuations generates tiny swings density, which grow in the process of inflation and become the embryos of future galaxies and stars. But it inevitably follows from this that the cosmic microwave background must preserve the memory of those events, a kind of "imprint" in the form of temperature fluctuations between different points of space. For a long time, it was not possible to measure this temperature spread - not enough sensitivity of the equipment. The breakthrough came in 1992 when the American satellite SOVE (Cosmic background explorer) and Russian "Relic-1" discovered temperature fluctuations background radiation. Them magnitude turned out extremely insignificant (temperature relic radiation is about 2.7 degrees Kelvin a deviations from middle not exceeded 0.00003 degrees Kelvin), that's why at all not marvelous, what before similar measurements were conjugated With considerable complexities. So or otherwise, but inflationary theory received reliable experimental the confirmation.

Start third millennium marked new achievements. After year and a half observations and analysis data received With help space observatories wmap, was presented much more detailed map distribution temperature of the cosmic microwave background radiation throughout the sky. English abbreviation MAP means Microwave Anisotropy probe, what can translate how "microwave anisotropic probe" (or probe), a letter W added in honour astrophysics Wilkinson which the was initiator project, but not survived before his endings. Except Togo, tar - in English "map". The value of Wilkinson's map is hard to overestimate. Analysis of received data and subsequent computer modeling allowed recreate picture birth and development of the Universe, to clarify its age and composition. This is a milestone event happened 13.7 billion years back (plus or minus 200 million years), what allowed put an end to the endless debate about when exactly the universe arose. Managed finally to find out, what space Universe geometrically flat, and exactly calculate one of the fundamental constants - the Hubble constant, which reflects the speed expansion of the universe. Judging according to the Wilkinson probe, this value is 71 kilometer in give me a sec on the one megaparsec distances (remember what one parsec - 3.26 light year). In other words, a one megaparsec (1 million parsec) area every give me a sec grows on 71 kilometers.

It has been established that the Universe, having cooled down after the Big Bang, remained for a long time dark and cold. First stars, on clarified data, started take shape through 400 million years after Big explosion, and so early them appearance extra once testifies in benefit of existence hidden masses (or dark matter), which their gravitational field collected smeared matter in lumps. Briefly speaking saying inflationary model showed myself reliable workable theory great consistent With experienced data. BUT therefore It has meaning take a closer look to her take a closer look following the stage stage history of our Universe.

By modern ideas, Universe is born in result random quantum fluctuations flitting out from singularities - dimensionless points, in which curvature space-time endless. Density substances in this point too reaches endlessly big quantities, a space and time apply in zero. In other words, neither space, nor time, nor matter in the usual sense in singularities not exists, a all famous laws stop work. Not It has there is no point in asking what was before, because before there was nothing: the singularity -this is ultimate the border, Rubicon, which the it is forbidden go. wanted would specially to emphasize that the described scenario of the birth of the Universe practically "out of nothing" is not empty fantasies of theoretical physicists out of the blue; it is based on rigorous scientific calculations.

The reader has already encountered the expression "quantum fluctuations" so many times that he must have been spinning on his tongue for a long time: what kind of animal is this and with what are eating? How, out of this accidental smallness, in fact, out of the void, can huge peace with planets, stars and galaxies?

People who are far from physics tend to believe that vacuum is the complete absence of something. whatever. Meanwhile, it necessarily follows from the theory of elementary particles that the physical vacuum is by no means emptiness, but the minimum energy of fields and particles, not equal to zero. It is literally stuffed with so-called virtual particles, which are born in pairs as if from nothing (for example, an electron and its antipode positron), from the heart frolic like mayflies and in a moment perish in an act of annihilation, leaving memory of oneself in the form of a quantum of light - a photon. Their lifetime is so short that it cannot to be measured in principle. Any measuring process limited natural physical limit - the speed of light, and virtual particles, emerging from the void, are destroyed So fast, what never not may be observed directly.

Incidentally, the fact that "empty" space cannot be completely empty With evidence follows from laws quantum mechanics. If a would vacuum was absolutely empty, this would mean that all fields (electromagnetic, gravitational, etc.) in it are in accuracy equal zero. However magnitude fields and speed his changes co time are analogous to the position and velocity of the particle, and the Heisenberg uncertainty principle, as known prohibits the simultaneous knowledge of both parameters: the more precisely one of the these quantities, the less exactly the second one is known. Not two peas per spoon - you have to to choose something one. Let's listen Stephen Hawking, famous Englishtheoretical physics:

...

Consequently, in empty space field not maybe have permanent zerovalues, So how then it had would and exact meaning (zero), and accurate speed changes (also zero). There must be some minimum uncertainty in field strength – quantum fluctuations. These fluctuations can be thought of as pairs particles Sveta or gravity, which in some moment time together arise, diverge, a after getting closer again and annihilate friend With friend.

...

Such particles are virtual ‹...›, in difference from real virtual particles cannot be observed with a real particle detector. But indirect effects produced virtual particles, for example small changes energy electron orbits in atoms can be measured, and the results agree remarkably well With theoretical predictions. Principle uncertainty predicts also Existence similar virtual steam particles matter, such how electrons or quarks. But in this case one member couples will be particle, a second - antiparticle (antiparticles Sveta and gravity - this is what same the very thing and particles).

However immediately arises question. Law conservation energy forbids her obtaining from nothing, and we, assuming the birth of particles from emptiness, this law seems to violated. casket opens simply. For start consider, how leads myself electric charge at birth couples electron - positron. Full charge remains equal to zero, since minus (electron charge) by plus (positron charge) in the end gives zero. Just for a very short time, the total zero charge is divided into two equal halves - positive and negative. Something similar going on and With particle energy: the electron has positive energy, and its antiparticle (positron) has, in some sense, equal quantity negative energy. So Thus, the total energy still remains zero at the moment of birth and subsequent mutual destruction virtual particles.

Similar considerations apply to the universe born out of nothing. For the first view, we are faced with an unresolvable paradox, because the part accessible to observations The universe contains an astronomical number of particles from which matter is built. Where did they all come from? The answer is simple: according to quantum theory, particles can be born from energy in form steam particle - antiparticle. Good, but where is taken breathtaking amount energy? Matter, filling universe (planets, stars and galaxies assembled from particles) has positive energy, but the world has also gravity, the energy of which is negative, so the total energy of the Universe is zero, as is its electric charge (the number of protons and electrons is the same). Butwhat available in mind when they say about negative energy gravity?

More once let's quote Hawking.

...

Matter in the universe is formed from positive energy. But all matter itself attracts itself under the influence of gravity. Two closely spaced pieces of matter have less energy than the same two pieces that are far apart, because that to spread them apart, you need to expend energy to overcome the gravitational power to unite them. Consequently, the energy of the gravitational field in some sense is negative. It can be shown that in the case of a Universe approximately homogeneous in space, this negative gravitational energy in accuracy compensates positive energy associated with matter. Therefore, the total energy of the universe is zero.

Very curious, what amount positive energy maybe double parallel doubling negative because the twice zero - all equals zero. At standard expansion this impossible, because the on measure increase Universe the energy density drops. But in the era of inflation, as we remember, the energy density false vacuum remains constant despite size increase Universe. Therefore, when doubling diameter our peace twice grow up and positive energy substances and negative energy gravity, a total energy Universe will be still dress zero. BUT because the in phase inflation dimensions Universe increase exponentially, on the orders orders then and general amount energy, required for education particle, too monstrously increases. Here to you, reader, and answer to the question of how, in such a miraculous way, all the matter that fills the Universe today, could fit in a tiny volume comparable to the Planck length. She is not there thought to put: when the inflaton field dropped to a minimum all stored in it potential energy gone on the birth elementary particles.

Let's get back to beginning began, to first moments life our peace, when he was just getting ready to fly out of the cosmological singularity. It must be said that singularity - very uncomfortable concept, because how abounds stockade very unpleasant infinity: endlessly small volume, endlessly large density, mass and temperature, the infinite curvature of space-time, and so on in the same vein. Physicists not by chance not love infinity, because what everywhere, where they appear, starts pandemonium: laws refuse work, formulas lose meaning, a consistent descriptions are coming apart at the seams. In that case, can you try get by at all without singularities, throw out them, So to tell, on the dump of history? After all talk goes about vanishingly small spatio-temporal scale, where classical physics Newton - Einstein already not works and where undividedly reign laws of quantum mechanics. Perhaps space and time, like charge, spin or magnetic the moment too have some limit divisibility, then there is, others words quantized? right whether we do this assumption?

And why not, after all? It is likely that in nature there is some indivisible cell space, his kind minimum distance, which not amenable to further crushing. If this is the case, then no body can collapse into a dimensionless point. Both the star and the universe as a whole will in this case collapse to a certain limit, until they hit an insurmountable boundary, and then inside a black hole there will be not a singularity with its tedious infinities, but peculiar quantum space, elementary volume diameter 10^{-33} centimeters. Because the overcome this is distance should one stroke (in otherwise case we turned out to be would in some moment in the middle indivisible segment, what impossible on definition), then there must also be a quantum of time - the minimum duration of any processes. A simple calculation shows that it is about 10^{-43} seconds, and both of these quantities, received titles Planckian length and Planck's time us already

Good known.

Planck quantities based on fundamental constants - the constant plank, speed Sveta and gravity permanent, - inevitably lead us to more alone important indicator - maximum possible density matter in null moment. Now she is already not endless although and unimaginably great - 1093g/cm3. This magnitude surpasses any imagination, for density atomic nuclei on the background these astronomical figures looks almost like an absolute vacuum. Suffice it to say that ten solar masses (and the Sun is a medium-sized star with a diameter of about 1.4 million kilometers) can easily fit in a volume quite comparable to the nucleus of an atom hydrogen. The temperature of such a superdense bunch also goes off scale beyond all conceivable limits and is approximately 1032 degrees Kelvin.

The meaning of the maximum possible Planck values is that no other parameters (smaller, when it comes to length and time intervals, and larger, if the conversation turns to indicators of density and temperature) cannot exist in principle. For example, ridiculous to ask, what happened through 10-45seconds after Big explosion, because the such moments time simply not It was. We reached began and ran into an impenetrable partition: there is no further way, for the usual concepts of space and time lose all meaning. In the area of Planck quantities missing subsequence events, there nothing not going on and because time nowhere flow. Space too loses connectedness, addressing in boiling chaos flashing and fading bubbles. We are unable to see this muddy brew, because the scales available to modern accelerators lie in the range of 10–16 centimeters, and on such distances space-time continues stay smooth. To personally behold Planckian scale, us had to would increase sensitivity equipment in 1017 times. And here then we saw would quantum ocean, abiding in able permanent chaotic seething, something like worried nautical an element that constantly drives wave after wave. However, from a great height, individual you can't see the waves - the ocean seems to be a calm water surface. And just going down lower we we can see succession fast running foamy lamb.

In the microcosm, at the level of Planck values, the space-time continuum collapses irrevocably, and space and time begin to foam. In this unusual world there is no certainty, there are no selected directions or sequence events, and therefore the American physicist J. A. Wheeler quite aptly called it quantum,or space-time, foam. Space and time become discrete, and concepts "before" or "later" lose any meaning. Derzhavinskaya river times broke into separate drops. And only when something suddenly emerges from nothing (random quantum fluctuation is experiencing a rapid inflation), the familiar to us are born space and time, and with them a new universe. Chaos gave birth to the Cosmos. So the way birth Universe identically birth space-time.

Justice for the sake of necessary Mark, what quantum character space-time is not the ultimate truth, but just a hypothesis, even if more or less convincing. Between topics long away not all scientists agree With such posing a question. Many physicists seriously doubt that spacetime and gravity is generally amenable to quantization: it is likely that these are purely classical objects. A business in volume, what birth Universe from quantum fluctuations (or space-time foam) must be described laws not existing on the today's day science - quantum theories gravity. However formulate these cunning laws, at least even at a theoretical level, so far no one has succeeded. it a task grandiose difficulties, and at all not by chance leading scientists put her on the first place among ten most difficult problems contemporary physics. M. AT. Sazhin writes:

...

General theory relativity (OTO) - relativistic theory gravity - fundamentally different from the theory of the electromagnetic field and the known fields of other types. GR connects the space-time geometry with the properties of matter. That's why construction quantum gravity equivalent to construction quantum geometry space-time. At this arises a lot of purely theoretical (quicker even formal mathematical) difficulties.

In other words, it is necessary to somehow link the quantum approach with the general theory relativity at description phenomena microworld. And to not confuse you, reader, finally, try short set out essence Problems, not going into in mathematical subtleties.

Quantum and classical approaches different fundamentally. At description movements particles classical physics operates notion her trajectories, then how quantum an approach insists Total only on the probabilities detection particles (in according to the uncertainty principle - the more accurately the particle velocity is calculated, the its location is less accurately known). In classical language, we say that an electron moves, but in quantum language you can't say that. It is more correct to say that the electron is in a certain state, described by a certain wave function, giving probability stay electron in volume or otherwise place. AT first case the equation movements is differential equation and easily is decided a in second requirement differentiability not performed. Mathematician will say what such probabilistic the trajectory is non-differentiable.

I will allow myself one more quotation from the book by M. V. Sazhin (if you, the reader, are not concerned formal calculations, you can easy to miss this paragraph):

...

So, in quantum mechanics trajectory is replaced notion probabilities find particle. AT theories fields concept particles is replaced notion quantities fields. It characterized by amplitude, phase and frequency. In quantum field theory, the amplitude, phase, and the frequency of any field are replaced by the concept of the probability of the same quantities. In general theory relativity role fields plays geometry space-time. AT her necessary work with the probability of having any geometry. But in general relativity the geometry must be differentiable, a in quantum gravity, how we seen on the example trajectories particle, this, in general saying not So!

It turns out, found braid on the stone. Theory relativity and quantum Mechanics stubbornly unwilling to join at the level of Planck values. And if ever they succeed consistent tie, then flow time in microworld will be be described peculiar wave function, denoting probability leaks some interval time although this is sounds, soft saying several unusual. However, the resolution of the paradoxes of quantum gravity may not be far off. One of the newest physical theories - So called theory superstrings - it seems, promises take off irremovable contradictions between quantum mechanics and general relativity. About this very curious theory we let's talk in next chapter.

In the meantime, to describe the birth of our world out of nothing, one has to involve the most general ideas about the quantum evolution of the Universe as a whole. At the same time, there must be several conditions. Firstly, to young fledgling Universe fluttered out from void without energy expenditure, its mass should be equal to zero. A little higher, I already wrote that positive energy matter compensated negative energy gravity, a because complete energy Universe (a means, and her weight) turns out equal zero. Laws saves in this case are not violated. The same is true for electrical charge. Finally, probability birth Universe from nothing calculated like under the barrier passage alpha particles in result process tunneling. What here available in mind?

When potential energy barrier a lot of above energy particle, she is, It would seem that in no case will he be able to overcome it. However, quantum fluctuations vacuum make revise this conclusion. Because the in compliance With principle uncertainty, the position and energy of a particle cannot be established equally accurately, we obliged accept in Attention quantum effects, inevitably influencing on the her behavior. Sooner or later, the energy of the particle will randomly increase abruptly and will become relatively large, as a result of which the potential barrier will be overcome. Like phenomenon movements over barriers known in physics how process tunneling. Something in the same vein once happened to our universe: although its complete energy was equal to zero, random quantum fluctuations allowed her tunnel in Existence from nothing.

So, emerging from space-time foam, newborn Universe for some time it swelled at superluminal speed (the theory of relativity, as we remember, it does not prohibit this, because it limits the speed of movement of material bodies), but when the energy of the inflaton field dropped to a minimum, there was a birth of matter in the form hot plasma. Inflation ended, replaced by the usual expansion, which we observe to this day.

The birth of the universe from quantum foam through a tunnel transition advocates theory eternal (or chaotic) inflation Andrew Linda. Of course term "eternal inflation" can not be interpreted literally. The inflationary stage is eternal exactly in that to the extent that, say, elementary particles are eternal, although each of them is born in his term is lost. Our Universe was in an inflationary phase of a quite finite (and very short) time, but universe one only our Universe not exhausted. There are a great many universes, they constantly emerge from space-time foam per check quantum fluctuations. This process randomchaotic and not It has neither the end neither start. Alone universes collapse, barely having time be born, others grow, remaining empty and dead, because the laws in them are such thatprohibit the emergence of complex structures, others turn into a kind of phantoms, for deprived time and development, a fourth are filled stars, galaxies and planets. By a happy coincidence, we live in just such a universe. Let's try explain mechanism eternal inflation on specific example.

At the Planck time (10^{-43} seconds), even before the start of inflation, physical processes managed spread maximum on the distance Planckian length (10^{-33} centimeters). Only in such an elementary volume by the beginning of inflation could there be reached thermodynamic equilibrium. However, the actual scale of the universe is not must necessarily be limited by the Planck length; it is likely that they were much larger and were a collection of tiny areas, each of which had size approximately equal to 10^{-33} centimeters. All these areas were isolated from each other. from a friend, because the light signal simply did not have enough time to penetrate from one areas in another. Consequently, physical terms in different areas noticeably differed, changing from region to region chaotically. Energy density inside elementary cells too significantly varied.

Let us once again recall the randomly scattered snowballs on the mountain slope: some lie nearly at the very the edges abyss, a other removed from her on the significant distance. AT In the vast majority of cases, the snowball rolls down unhindered and easily reaches points minimum. AT such "prosperous" areas inflation ends relatively quickly (as we remember, it continues as long as the snowball is on the plateau) and is replaced banal extension on law Friedman - Hubble. But painting complicated by the fact that individual lumps under the influence of random quantum fluctuations may move and in directly opposite side, reaching unimaginable speeds, because the process inflation develops on exponent. AT such areas inflation not will never end.

To visualize this in any way, imagine a rubber sheet or plastic film, lined into cells like a chessboard. Each of fields, relevant in given case elementary Planckian volume, can stretch how whatever strongly or, vice versa, leave in immunity. AT resultwe we get confused conglomerate, consisting from fragments unified the whole deformed purely individually. "Calm" plot, where inflation a long time ago ordered for a long time live, maybe to be surrounded by countless quantity regions, located in absolutely different modes: in some inflation straightaway same choked a in others continues until now since.

That's why the size observable now Universe (Metagalactics), component 1028 centimeters, which roughly corresponds to 10 billion light years, may bean insignificant part of the universe as a whole. There, beyond the event horizon, others live and live worlds that have nothing to do with our universe. And although they are formally associated with her indisputable fact commonality origin, With physical points vision they are

"things in themselves", because they have nothing to do with our Universe. Scenario of eternal stochastic (probabilistic) inflation describes all possible universes, which in famous sense exist "somewhere" in space.

BUT. BUT. Starobinsky, corresponding member RAS and main scientific employee Institute theoretical physics them. L. D. Landau, given simple question:

...

What is the practical significance of all this? We can't see these other universestherefore, this does not lead to new observational effects (or we have not yet learned how to find - it should be recognized that the whole theoretical picture of the Metaverse is not yet developed). However, from an ideological point of view, it is clear that all the hot previous discussions about the "one-time birth of the universe" were naive. It became clear that our the visible universe is only one of the possible realizations of universes that are constantlyoccur in the Metaverse in different places in space (and even in some sense in different times - time in other universes, generally speaking, does not have to correlate with time in our universe).

Short sum up said. Birth classical space-time from quantum foam was the result of a random quantum fluctuation, and the age of the universe was then about 10-43 seconds. Diameter Universe in that time was a little more 10-33cm, a density this microscopic clot reached monstrous values - 1093 g / cm2 (the so-called Planck density, the maximum possible in nature). Temperature too was under become - near 1032 degrees Kelvin. AT progress inflation, duration which was several Planck times (10-43–10-37 seconds), the temperature varied over a very wide range, rapidly falling to zero. Swift inflation smoothed out space and did his practically homogeneous in all directions. The era of inflation is basically a cold stage; elementary particles more No, a matter presented scalar inflatonic field.

When inflatonic field reached minimum potential energy, happened the birth of matter in the form of hot plasma from quarks, gluons, electrons and their antiparticles. The Universe warmed up again to very high temperatures of the order of 1026-1029 degrees Kelvin. Exponential inflation has been replaced by the usual leisurely expansion on law Hubble, what perceived us how Big explosion. Early Universe represented yourself his kind hot quark soup: high temperature prevented their unification, and therefore each quark lived an independent life. By measure fall temperature they started unite in nucleons, So how Existence quarks in the form of free particles at relatively low temperatures is impossible. When The universe cooled down to about 1011-1012 degrees Kelvin (its age at that time was 10-4 seconds), there are no free quarks left in nature - they all united into protons and neutrons. This process received call baryosynthesis, or quarkadron phase transition. By this time, the space of the young Universe had turned into a thick mess of protons, neutrons, electrons, neutrinos and photons, as well as their antiparticles. However, here which is curious: if particles and antiparticles were equal at the end of inflation, then they inevitably must were would mutually be destroyed in process annihilation, and then building material necessary for the formation of stars, galaxies and us, simply wouldn't be enough. In other words, why did symmetry breaking happen? between particles and antiparticles?

So, laws nature are the same for particles and antiparticles, a because not bad would figure out, how arose baryonic excess. On the any happening note what final response on the this question No, available several versions more or less convincing, and each of them requires the involvement of a complex mathematical apparatus. That's why confine ourselves to a simplified model, which, but, helps understand essence affairs.

Let us introduce a hypothetical field that interacts equally with both particles and antiparticles, and denote it by the Greek letter 0. We represent it graphically, in the form parabolas. The field energy will be maximum on its branches and minimum in the bottom area, in point, lying on the axes abscissa. For visibility can introduce yourself pit or some kind of vessel, say, a bowl or a wine glass widening upwards with a rounded bottom. We place a ball on the inner wall of the bowl and assume that its energy is the greater, the above it is located. Rolling down to the bottom bowls, the ball loses energy.

Now remember that at the time of the birth of our Universe, the energy density was very great. AT further she is all time fell striving to zero, a energy fields passed in energy born particles. AT our models antiparticles must to be a little less. But how this achieve? Suppose what particles born at the field moves along the left side of the parabola, and the antiparticles - along the right. The painting continues remain completely symmetrical: neither particles nor their antipodal twins have even account no benefits, because the quantum fluctuation - embryo our Universe
— with equal probability can occur both on the left and on the right branch. And now let's see, what happens next.

About this Good and simply tells FROM. G. Ruby:

...

The moment of truth comes precisely at the birth of our universe. If we live in Universe, accidentally born on the left branch, then the following happened. The field starts move down and spawn particles. Then it "skips" the position of the minimum and climbs to the right branch of the parabola, but part of its energy has already been given to particles, and it will rise below elementary values. That's why, when starts traffic back to minimum potential energy, the field generates antiparticles in smaller quantities. These fading fluctuations continue enough for a long time, and total amount particle, certainly, not will be coincide With quantity antiparticles - simply because, what their By birth on the left branch of the potential, the Universe broke the symmetry of the theory. This is exactly what we were looking for! By the way, if the Universe were accidentally born on the right branch, then we would be dominated by antiparticles. We would be composed of antiparticles, but of course we would call would them "particles".

And dark came

The wind brought us comfort
And in the azure we felt
Assyrian dragonfly wings, Busts
cranked darkness.

Osip Mandelstam

Previous chapter nearly entirely was is devoted distant past our Universe. The picture that emerges is strange, absurd and a little frightening: a huge world, inhabited countless many stars and galaxies, arose literally from nothing, practically from emptiness, from some insignificant quantum fluctuations. However and in the modern state of the Universe is also full of oddities, and the first place among them in right belongs to the mystery of the hidden mass, which is also called dark matter, and dark energy (not confused with hidden mass).

Observations of the last two decades have shown that the fraction of the ordinary visible substances - protons, neutrons electrons and photons - account for not more four % gravity mass-energy Universe (then there is mass-energy, creating gravitational field). Rest 96% - this is some enigmatic substance, which not radiates and not absorbs Sveta, a her presence can discover only only on created her gravitational field. She is no way not interacts With ordinary matter, so the epithet "dark" should be recognized as not entirely successful: with the same success it could be called "transparent" or "invisible". In other words, majestic a round dance of celestial bodies, which meticulous astronomers have studied for centuries, in fact turned out to be an insignificant surface part of an iceberg resting on an invisible dark block unknown what. About the physical nature of this incorporeal, but very weighty ghost contemporary the science not maybe to tell nothing certain. More Togo, at all recently it turned out that the dark underside of our world is heterogeneous and breaks up, in turn, into two components that are very different in their properties: dark matter (it is also hidden mass), which is approximately 25% of the total mass-energy, and dark energy (71%). However about everything in order.

The first bell, indicating that all is not well in the Danish kingdom, rang in 1933 year, when American astronomer of Swiss origin Fritz Zwicky conceived measure complete mass groups galaxies on them luminosity. He I did it simply: I counted the number of stars in each galaxy and multiplied this number by middle mass stars. It seemed would, reliable and verified method. However another an approach, founded on the law world gravity and evaluation speeds stars, gave incomparably large amount of mass. Zwicky noticed extremely curious anomalies in movement individual galaxies inside clusters. Any by chance taken galaxy moved in such a way as if the total mass of the cluster greatly exceeded the sum masses of its constituent galaxies. Since this hefty "appendage" is invisible and can be discovered only by the nature of gravitational indignation, Zwicky proposed to name his dark matter.

At that time, the scientific community reacted to Zwicky's proposal quite sluggishly, and only 40 years later they started talking about the hidden mass again. In the 70s of the last century anomalies, similar topics what kind discovered American astronomer, were revealed in spiral galaxies. As you know, spiral galaxies, unlike galaxies of another types (elliptical and irregular) rotate, however this rotation has nothing in common with the rotation of a children's spinning top or top. The galaxy is not solid body, a consists from dozens billion stars, each from which moving herself on yourself describing closed curve around galactic center. From here follows, what in in accordance with the laws of celestial mechanics, the speed of a star as it moves away from the center should fall. In any case, the planets of the solar system behave exactly like this: farther planet lags behind from the sun, topics below her orbital speed.

BUT here traffic stars in spiral galaxies on incomprehensible reason this immutable law not obeys. Astronomical observations testify about volume, what speed all stars, beginning With some distances from center, becomes a constant value. How to resolve this unpleasant situation? Put my hand on my heart we have little choice. One of two things: either the masses of galaxies are estimated incorrectly, or the laws Newton's principles are not universal and can be violated under certain conditions. Second option looks too extravagant and is not seriously considered by most scientists, although separate heretics from physics allow such possibility. Let's say Israeli M. Milgrom relatively recently proposed hypothesis received title modified Newtonian speakers (MOND). According to this hypothesis traffic stars, clouds of interstellar gas and other objects in the outer layers of spiral galaxies obeys not Newton's law, but a more general law, which includes Newtonian mechanics how private happening. Accelerated traffic stars explained topics what on the big distances from the galactic center, Newton's usual law does not hold, because strength gravity acquires a different size.

Tem not less majority specialists point vision Milgrom not share. The modified dynamics not only sins with a lot of frank exaggerations, but also does not agree well with the data of observational astronomy (for example, it is unable to explain character movements substances in clusters galaxies). That's why nearly all astrophysicists tend to explain anomalies in the motion of stars by the presence of invisible (dark) matter, which, like a huge spherical cloud, envelops every galaxy. Calculations show that in the case of our galaxy, the diameter of such a halo must be at least 300 thousand light years, then there is in three times surpasses diameter of the Milky Ways.

But what is the physical nature of this unusual substance, which, as we remember, it accounts for 25% - more than six times more than ordinary matter, emitting light? First, candidates for the role of dark mass carriers can be compact bodies, the so-called massive astrophysical compact objects in the halo Galaxies - Massive Astrophysical Compact Halo Objects (MACHO). Among these dark formations relate black holes, brown dwarfs, old neutron stars, clouds of weakly interacting particles and possibly white dwarfs. All of them should not glow, otherwise they would have been discovered a long time ago. Brown dwarfs are something average between gas giant planets and small light stars. Weight such object not must exceed ten % masses sun, otherwise inside him flare up thermonuclear reactions that will lead to light emission. black holes and neutron stars, claiming on the role compact objects, too must satisfy certain conditions. The former have no right to be too massive, since the radiation from the matter falling on them will immediately give them away, and the latter must have a very respectable age, since only old neutron stars practically not radiate and because invisible.

Under the influence of gravitational forces, dark matter is distributed unevenly, simply saying crowded, like ordinary matter, and astronomers study character this distribution various methods - on crooked rotation galaxies, them large-scale structure, gravitational lensing, and so on. Under the last the occurrence of false images is understood, since the gravitational fields of the hidden mass distort trajectory movements Sveta from distant sources. However observations show what some only compact objects clearly not enough for successful permissions Problems dark matter. That's why physics, involved in study elementary particle, believe what phenomenon hidden masses tied in first turn With So called WIMP - Weakly Interacting massive Particles (weakly interacting massive particles). These hypothetical particles bye not discovered, and then circumstance, what they extremely weakly interact With substance creates large difficulties in proving their existence. Such particles are sometimes called cold, or non-relativistic dark matter, because they are moving co speeds, a lot of smaller how speed Sveta. However them slowness With overwhelmed bathe very decent weight, because the mass of weakly interacting particles is 1000 or more times surpasses mass atom hydrogen.

By the way, in addition to the cold in the Universe, there is also hot dark matter in the form relic neutrinos with nonzero rest mass, but their contribution to the total gravitational mass-energy not exceeds one and a half percent. How we we see work at astrophysicists still no end, but to doubt the real existence of dark matter today is already not have to because the exactly she is contributes main contribution in mass galaxies.

But more more mysterious properties has dark energy, on the share which accounts for 71% of the total mass-energy of the universe. Unlike the hidden mass, it is not crowds under the influence of gravity, but strictly uniformly and uniformly fills everything space universe, like ideal solid environment, and everywhere and always It has constant density. The dark energy hypothesis (which, strictly speaking, has now become full theory) emerged in 1998 when two international teams of astronomers announced the discovery of the accelerated expansion of the universe. This fundamental fact meaning whom difficult overestimate, was installed at observations per distant supernovae stars certain type (type 1a). Such supernovae have exclusively high luminosity, comparable co luminosity whole galaxies, in which they flare, and therefore are clearly visible at intergalactic distances. Except Moreover, a unique feature of type 1a supernovae is the fact that their own the luminosity at maximum brightness lies within very narrow limits. In other words, power radiation stars this type practically identical, and because them received call
"standard candles." From school course physics known what flow light radiation decreases back proportionately square distances from source. So way, measuring the brightness of a supernova on Earth that erupted in a distant galaxy, and comparing it with the real intrinsic luminosity of the source (which is known), one can calculate distance before object. Especially important outbreaks supernovae type 1a in very distant galaxies, since cosmological effects become significant and one can not only define permanent Hubble, but and measure parameter density universe, then there is install her geometry.

Observational data on supernovae of type 1a, accumulated to date, allow us to state with a probability of 99% that the Universe is expanding at an accelerated rate. And it is very curious that the mode of the standard Hubble expansion did not change yesterday and not today, but at least several billion years ago. It is difficult to name the exact date but if believe archival photographs stellar sky, most remote from us
"standard candle" lit on the distance in ten billion light years from planets Earth. Its luminosity fits perfectly into the parameters of the Friedmann model, which implies to conclude, what more ten billion years to that back Universe continued expand classically - in complete compliance With law Hubble. However character shine more young supernovae does not allow to doubt that 7–8 billion years ago the dark energy prevailed above forces gravity and Universe became expand faster.

Builds up impression, what dynamics universe governs some
"expanding" field. As long as the volume of the universe is relatively small, gravity is effectively counteract the expansion of space, but sooner or later there comes a moment when when the density of matter falls below a certain critical value and the field, density which does not change over time, begins to inflate space more and more energetically. More Togo, pace extensions turns out in accuracy such what makes recall

the notorious "lambda", the cosmological constant that Einstein introduced into the equations general theory of relativity back in 1917. Einstein's universe was static, and he needed the lambda member in order to balance the constricting force of gravity universal cosmological repulsion: otherwise, all matter must inevitably gather together. Einstein himself could not stand his "lambda" and subsequently called the introduction of the lambda member "life's biggest mistake". However, after in 1922–1924 Leningrad mathematician A. A. Fridman found a non-stationary solution Einstein's equations, and the American astronomer Edwin Hubble in 1929 discovered red bias in spectra distant galaxies, became clear, what Universe With moment his birth is constantly evolving, and the inconvenient "lambda" has been safely forgotten. Oblivion stretched for more than 40 years, and only at the turn of the 60s - 70s of the past century about cosmological constant started talking again. From works domestic theoretical physicists E. B. Gleaner, BUT. BUT. Starobinsky, I. B. Zeldovich and someothers followed that the vacuum can have non-zero energy. In this case, the hypothesis cosmological constant is equivalent to the idea of a perfectly homogeneous medium, evenly filling all the universe. Properties such environments very unusual: her pressure expressed negative size, a density unchanged in time and space. And as soon as the pressure is negative, then at a constant density it will be create anti-gravity Effect, accelerating extension Universe. Therefore, quite probably, what dark energy there is not what other how manifestation vacuum fields With negative pressure.

Does this remind you of anything, reader? Then go back to the beginning of the last chapter, in which speech walked about cosmological inflation - period ultrafast extensions newborn Universe. Hypothetical inflatonic field, effectively inflated space near points "zero", had exactly such same characteristics - extremely strong negative pressure and a constant density that does not change with time. Therefore, we have the right to assume that the inflaton field has not gone away, but continues to be present in our universe. Then dark energy will be just such a field, located in minimum his potential. Between by the way, from here follows important consequence: era inflation qualitatively absolutely similar the one to which our The universe is approaching today. Undoubtedly, there is a difference between them, but it is purely quantitative character. It is clear that at the dawn of history, in the stage of inflating all the meanings curvature of space-time and effective energy density were in a colossal times more than now, but there are no fundamental differences between these two eras seen.

So, until 1998, it was possible to speak with confidence about the three components of matter, evenly filling space Universe. Firstly, this is usual substance -protons, neutrons, and electrons that make up stars, planets, and as little aswe With you. Secondly, this is mysterious dark matter (hidden weight), consisting fromnon-relativistic particle, not radiating Sveta and practically not interacting With ordinary substance. Finally, thirdly, this is the "residual" radiation - relic photons and neutrinos, preserved as an echo of the hot beginning of our world. Not discovered until now, gravitons and some other ultrarelativistic particles also fall into this category. These three incarnations universe provide worldwide gravity, a here fourth component, on the share which account for two thirds complete density contemporary universe, identified at all recently and creates phenomenon universal cosmological repulsion. So the fate of the world is controlled by a certain continuum with positive constant density and negative pressure, and in absolute expression these two quantities equal between yourself.

Regarding the physical nature of this mysterious substance, we are currently day we can not say almost anything. If interpreted as a kind of cosmological permanent, we inevitably we balk in jewelry accuracy initial parameters, that most thin setting, which for a long time imposed in teeth. It turns out, what initial the potential energy of the universe was calculated so flawlessly that as subsequent "calm" expansion managed to provide such a critical density our peace, which did space nearly perfect flat. "Why anti-gravity action dark energy appeared only at that time, when become arise galaxies? - ask some astrophysics. Truth, these discrepancies removed in the scenario of chaotic inflation A. D. Linde: cosmological constant maybe accept various values, and only there, where exist stars, galaxies and in general, complex structures, it acquires such a value that allows the appearance questioning subject. In other words, dark energy is unevenly distributed in space, a because version divine fishing can co calm soul close. AT those corners universe, where meaning cosmological constant on will blind chance turned out to be different, asking about the jewelry fitting of parameters is simply no one.

Meanwhile, not all physicists are ready to agree with such a formulation of the question and It is believed that the density of dark energy is not of a vacuum nature and may eventually change. Let's say the Americans Paul Steinhardt and Richard Caldwell think that under the mask dark energy hiding special quantum field, which maybe accept variables values. In memory of ancient thinkers, they called it the quintessence. As is known, the ancients believed that the components of the universe are four elements - earth, water, fire and air, but restless Aristotle added this nomenclature fifth essence - the quintessence of which the etheric bodies supposedly consist. In the disputes of highbrow theorists we will not interfere, but we will only note that the question of the physical nature of dark energy still very far away from final approval. So or otherwise, but the leading role dark energy in the evolution of the universe in our days of doubt no longer calls. What would she neither was at the microscopic level - a special vacuum energy or geometric radical invested in the universe - but the fact remains: for several For billions of years, our Universe has been expanding at an accelerating rate, and the tone for this expansion is set by exactly dark energy - some substance With negative pressure and constant density.

Based on the foregoing, the entire history of the universe can be divided into four epochs and describe with a four-term formula of the following form: ... DS (I) - FI - FM - DS ... First the link of this formula denotes the phase of inflation (the letter "I" in brackets), and the combination "DS" indicates the de Sitter character of the expansion. Although about the Dutch astronomer Willem Sitter we have already mentioned, it is necessary to make a small explanation. He was one of the first scientists to recognize the general theory of relativity, but the stationary model Einstein did not suit him. Einstein's universe was described by Riemannian geometry and was a four-dimensional hypersphere, an analogue of which in three dimensions can bebe the surface of a rubber bladder or a balloon. This universe is closed itself and has no boundaries, although its scope is finite. A beam of light, if it meets no obstacles, would propagate in such a model along a circle (more precisely, along a geodesic line, because the shortest through between two dots on the surfaces spheres is exactly such curve).

Sitter proposed dynamic model empty and continuously expanding universe, similar on the air ball, which the all time inflate. By measure inflation diameter ball constantly growing, a his geometry, continuing stay Riemannian all more and more approaching to geometry Euclid. Others words space in such a universe becomes more and more flat, and the beam of light does not move along circles, but in a continuously expanding spiral. However, Sitter was very unlucky. He was too far ahead of his time, and his hypothesis remained in the memory of his contemporaries. graceful and witty mathematical incident. Universe Sitter expanded on exhibitor (then eat in geometric progressions in dependencies from time), what in that time (in

1917) contradicted the observations. But proposed a few years later model BUT. BUT. Friedman insisted on the volume, what objects are removed friend from friend co speed directly proportional to the distance up to them.

Today we understand that this contradiction is imaginary. And Friedman was no fool, and Sitter too not bast shoes cabbage soup slurped: each was in my own way rights. AT era inflation the space grew exponentially - in full accordance with Sitter's calculations. BUT when the energy of the field bursting the Universe fell to a minimum, the expansion mode immediately same changed. And on the stages radiation (FI-phase), when Universe was red-hot clot hot plasma, and on the stages recombination (FM phase), when radiation separated from matter, our world expanded proportionally - according to Friedman's law - Hubble. But when the Universe grew considerably and cooled down, dark energy again entered into your rights. Several billion years ago, the era of the dominance of the dark energy, which continues before now since, and Universe again start expand accelerated. BUT because the on their dynamic parameters contemporary era nearly nothing not is different from stages inflation, BUT. BUT. Starobinsky proposed name her de Sitter (abbreviation DS in the right side of the formula).

Incidentally, the problem of the dark energy has very curious philosophical aspect. Until the moment when the force of the universal cosmological repulsion became dominant a Universe start expand accelerated managed happen a lot of different events. Before entering the mode of accelerated expansion, the world went through an era inflation (DS(I) - stage), radiation phase (FI-stage) and phase dominance dark matter (FM-stage), when the radiation is separated from the matter. Hence we have full right to assume that the inflation phase on the left side of the formula was preceded by some developments.

BUT. BUT. Starobinsky writes:

...

All four stages and transitions between them, included in this formula, may to be calculated in theory and explored on existing observant data. However, is it possible to think that this chain contains the entire evolution of our Universe in past and future? I guess what no. How once vice versa, wonderful quality analogy between DS(I) - and DS stages, explained above, suggests us, what this chain
– just a small piece of something much larger, perhaps even infinite. Let's look along the formula from right to left. We see that before the DS stage there was a long and varied backstory. Then it is natural to expect that the DS(I) stage also had its own background (ellipsis to the left of the formula). Now let's look from left to right. It's obvious that DS(I) - the stage was unstable, the primary dark energy decayed into others (including including ordinary) types of matter. Why then does modern dark energy have to be stable and cannot transform into other types of matter in the future (ellipsis on the right from formulas)?

Of course, the duration DS stages many times greater than phase of inflation because quantum systems with lower total energy are much more stable. What concerns pre-inflationary stories our peace, then majority contemporary cosmological models forbid ellipsis left from formulas and insist on the occurrence Universe from nothing (from nothing). However, on opinion BUT. BUT. Starobinsky, there are countless other scenarios in which DS(I) - the stage is preceded by something. He writes that together with Ya. B. Zeldovich they formulated the opposite concept of the birth of the Universe "from anything" (from anything), but, due to extreme her extremism not considers her in detail. One word, attempts to know, what preceded phase inflation, not stop, and to be maybe, us waiting on the this way more a lot of interesting discoveries. So or otherwise, but world turned out to be immeasurably harder how seemed scientists more any thirty years back.

And what about the distant future of our universe? What is the coming age to us trains? There are several answers to this question, because the physical nature of the dark energy is still a mystery with seven seals. In the simplest case, if the vacuum energy is positive and does not change with time, the universe will expand indefinitely. The night sky will begin to empty little by little as more and more objects move beyond horizon of events, and in 10-20 billion years at the disposal of mankind will remain our Galaxy (Milky Path), neighboring nebula Andromedae Yes more several galaxies from the so-called Local Group. After 10^{14} years, new ones will cease to be born stars and in the universe there will be only bodies that almost do not give light - white and brown dwarfs, neutron stars and black holes. But in the end all the stars will go out and die, and in 10^{37} years in an exorbitantly swollen space it will be impossible to find anything but black holes and elementary particles. But nothing is forever. Due to quantum processes, black holes after all radiate, although and very slowly, a because early or late they too evaporate. it event will happen when age Universe will be 10^{100} years and all universe will be filled extremely sparse gas from stable elementary particles - electrons, three kinds of neutrinos and, possibly, protons. Peace again will become empty, how biblical Earth in early began, because the distance between two particles will be far surpass dimensions contemporary Universe.

What and say, a heartbreaking sight. However, these are still flowers, because there are far more catastrophic scenarios for our distant future. One of them shows what in world generally nothing not will remain. A business in volume, what if usual extension Universe in form continuous growth her space not generates no forces acting on physical bodies, then dark energy behaves completely otherwise. Accelerated inflation likewise appearance some strength, stretching all objects. Today, its magnitude is vanishingly small - 10^{30} times weaker than gravity on the surface Earth. If the acceleration grows steadily exponentially, then, in the end, the matter will end not only with the destruction of all physical bodies, but even elementary particles, from which all matter is built. The universe will turn into a swelling nothingness, empty into in the most literal sense of the word. This pattern, called the Big Gap rip in English), was suggested in 2003 year in article R. R. Caldwell, M. Kamionkovsky and H. H. Weinberg "Phantom energy and space the end Sveta". However, not everything is so hopeless: other astrophysicists, for example, Stephen Hawking believe that expansion will sooner or later be replaced by contraction. Frankly speaking, such a prospect also does not bode well for humanity, but this is already a separate song.

However, the coming years lurk in the mist, as the classic once wrote, and therefore not let's guess on the coffee grounds, but we will turn our faces to the past. In the previous chapter the theory of superstrings was mentioned, which seems to be consistent quantum mechanics and general relativity. It's time to talk about her in more detail, especially since string theories in various versions are very popular today and very lively discussed.

For start remember about four types fundamental interactions - electromagnetic, strong, weak and gravitational, under the sign of which this imperfect world. Briefly let me remind you to you, reader, what electromagnetism was exhaustively described by the English physicist James Maxwell in 1873. If not this strength, built on the confrontation two polar began (charges one sign repel each other, and opposite ones attract), neither atoms nor molecules could exist. Chemistry and biology So or otherwise come down to electromagnetic interaction. TV and radio, thanks to which we learn about tsunami in indonesia, escapades unfinished Taliban in foothills Hindu Kush or next takeoff prices on the oil on the world markets, too obliged their existence phenomenon electromagnetism.

strong interaction holds protons and neutrons inside atomic core, counteracting the forces of Coulomb repulsion, and also glues together subnuclear particles are quarks, from which all matter is built. Weak interaction (weaker than it only gravitational) answers per transformation elementary particles in microworld and some kinds radioactive decay.

Finally, gravitational interaction (it most weak from all - the electromagnetic repulsion of opposite charges exceeds the contracting force gravity in 10^{43} times) compels body be attracted friend to friend and It has only one sign
– mass (what is "mass" and where it comes from, no one knows). But electromagnetic forces operate only on the charged objects, a gravity - on the all body without exceptions, having mass. And since macroscopic structures are almost always electrically neutral strength world gravity acquires defining role in cosmological scales.

carriers electromagnetic interactions are photons (if more precisely, virtual photons), strong - gluons (from English adhesive - "glue", "glue") weak
– So called heavy vector bosons (W^+-boson, W^--boson and Z^0-boson). A here gravity costs in this row apart, because what carrier gravitational interactions - hypothetical graviton - before now since not discovered. That's why gravitational field described in framework general theories relativity how curved four-dimensional space-time continuum. Curvature space is determined by the presence of masses, and these masses themselves, as already mentioned before, they do not move in a straight line, but along trajectories of the smallest length - geodesic lines. Let's remember simple example. If a put on the elastic rubber sheet weighty metal ball, rubber sag, forming hole. If a now take ball slightly less and try to roll it past a heavy ball, it will either roll into a recess (it will be attracted to heavy ball), or describe near him some curve what will be depend from speed lung ball and distances between them. How more weight, topics stronger warps space. Others words strength gravity is equivalent to bend space-time.

It remains to add that electromagnetism and gravity are long-range forces, while the strong and weak interactions are effective only at small and ultrasmall distances (10^{-13}– 10^{-15} cm and 10^{-16}– 10^{-17} cm respectively).

AT 1967 year in physics elementary particles happened significant event. American Stephen Weinberg and Englishman Abdus salam regardless friend from friend showed that the electromagnetic and weak interactions are of the same nature and have a common origin. Separately, they act only at relatively low temperatures, and at temperature order 10^{15} degrees become indistinguishable uniting in electroweak force. From the Weinberg-Salam model it followed that in addition to the photon there are three more particles that are carriers of the weak interaction, - vector bosons already familiar to us ("double-ve plus", "double-ve minus" and "z zero"). At high levels energy, relevant temperature 10^{15} degrees Kelvin (a temperature, as is known, is only a measure of the amount of energy), W-- and Z-particles begin to behave exactly like a massless photon. This is similar to the behavior of the ball when playing. in roulette. Stephen Hawking writes:

...

At high energies (then there is at rapid rotation wheels) ball leads myself nearly equally - non-stop rotates. But when wheel slow down energy ball decreases and in end ends he fails in one from thirty seven grooves, available on the wheel. In other words, at low energies the ball can exist in thirty-seven states. If for some reason we could observe the ball only when low energies then considered would, what exists thirty seven different types balls!

ten years later theoretical model Weinberg - Salama brilliantly confirmed experimentally: three types of heavy vector bosons were found, and it is with the parameters predicted. The success exceeded all expectations, and today law counts, what significance models Weinberg - Salama, received title standard model, is quite comparable with the achievements of the great Maxwell, who combined in his time electricity and magnetism.

But if electromagnetism and weak forces are two sides of the same coin, then maybe the strong interaction is nothing but a kind of some common force? And indeed, the standard model predicts that at even higher temperatures (near 1028 degrees) must happen an association strong and electroweak interactions. Photons, gluons and vector bosons begin to behave identically and they all become "on one face", like the three hypostases of the Creator - God the father, God the son and God the spirit St. carrier this universal interactions must to be mysterious Higgs particle (or X-boson), which has not yet been experimentally detected. However physics not lose hope, what Big hadronic collider - largest in world elementary particle accelerator built on the shores of Lake Geneva and launched in the fall of 2007, will help to dot the i's. Incidentally, the Higgs boson notable more and that which gives weighing everything rest particles.

So three interactions out of four – electromagnetic, strong and weak - at certain conditions merge together before complete indistinguishability. Such terms existed in the very early universe, when its age was estimated to be microscopic fractions of a second. First, the strong interaction separated from the common trunk, and then electroweak, which, in turn, as the temperature fell, decayed into weak and electromagnetic. A theory that claims to unite all three forces (it, alas, has not yet been built), called theory of the Great associations.

BUT how to be With gravity? Logics suggests what at temperatures order 1032 degrees, it must inevitably merge into a triple union, turning a truncated trio into a full quartet. The catch, however, is that if three forces within a quantum mechanics without special labor unite in single strength (on extreme least purely in theory), then gravity in this formula not climbs, stubbornly not wanting succumb quantization. She continues to be the fifth wheel in the cart, and when trying to combine quantum approach with the general theory of relativity from all cracks immediately begin crawl out ridiculous infinity. So what epithet "great" with regard to to theories the unification of three forces sins with a certain stretch: to squeeze gravity into the Procrustean bed hypothetical unified superpowers no way not succeeds.

Between topics way, allowing consistent tie gravity With electromagnetism, was proposed more in early of the past century (about two others interactions - strong and weak - at that time they did not know anything). In 1919 the mathematician Theodore Kaluza wrote Einstein letter, in which detail outlined my idea unification of electromagnetic and gravitational forces. As you know, Einstein's theory formulated in framework representation about four-dimensional space-time (three spatial measurements a plus one temporary). Kaluza proposed enter additional spatial measurement and built model five-dimensional space-time (four spatial dimensions plus one time), and was able to show that his five-dimensional model is identical to Einstein's four-dimensional model a plus electromagnetism. Others words in theories Kalutsy fifth measurement space
"answered" per electromagnetism: he proved what introduction additional

spatial measurements equivalent to introduction electromagnetism.

According to Einstein, gravity, as we remember, is a manifestation of the four-dimensional metric spacetime, a Kaluza found non-quantum, geometric solution for electromagnetism. It followed from his theory that gravity in the world of five dimensions is one, and in four-dimensional space-time Einstein she is speaks in form two forces - gravity and electromagnetic.

Kaluza's model was flawless from a mathematical point of view, but contained significant inconsistency. He failed to explain why the fifth dimension of space does not manifest itself in any way in our real four-dimensional world. We will try to eliminate this space, using to a simple analogy.

Any cord, rope or hose is, without a doubt, a three-dimensional body - cylinder. If a we we will consider such cylinder With enough big distance, then its length will come to the fore first of all, since the other two measurements (height and width) are much inferior to it in size. Look at the human hair or web thread: these are exactly the same cylinders as a thick rope, but two measurements due to their smallness are practically not perceived by us. Spider web or hair look like a one-dimensional line.

It is quite possible that the space of our Universe is organized similarly: three spatial measurements stretched out before cosmological scale, a fourth so little that "caught" even with the help of the most sensitive laboratory technique, not to mention seeing it with the naked eye. We cannot see the fourth dimension of the space of our Universe for exactly the same reason as not in able see additional measurements the thinnest threads. But staying fundamentally unobservable, it nevertheless manifests itself on a large scale as a force electromagnetism.

Ideas Kalutsy were developed in 20s years of the past century Swedish mathematician Oscar Klein and got title theories Kalutsy - Klein. Long time they presented themselves speculative speculation not having relations to real physical world, but these days have become quite popular. The point is that if electromagnetism maybe to be explained involving additional measurements space, then is it possible to do the same with other types of universal interactions - strong and weak? Perhaps they are also connected with some hidden dimensions beyond our perception. Then the picture of the universe immediately simplifies, acquiring a slender and finished look. Let's call these compact hidden measurements by *internal* space, and the three large dimensions - *outer space.* If a structure external space determined forces gravity, then the form internal space will be tied With three others interactions - weak, strong and electromagnetic. It is clear that such a single description of all the forces of nature on language geometry appears very attractive.

However, two very serious questions must first be answered. Question one: how the interior space is arranged, how does it look upon closer inspection? Question second: if the Universe is multidimensional, then why are there only three spatial dimensions puffed up to cosmological scale?

Let's figure it out on order. Firstly, internal space must to be very small. By all probability, his the size lies in areas Planck lengths (near $10-33$cm). Secondly, despite its smallness, it should not have boundaries. Otherwise case, elementary particles, having reached the edge, would behave in exactly the same way as balls on table top: they would roll down. Therefore, the interior space to be simultaneously and compact, and rolled up, then there is closed self on the myself. Finally, remember about volume, what curvature space (in given case speech goes about external space) is closely related to gravity. If the inner space It was too twisted, this is caused would additional gravitational effects. BUT because the we them not we observe remains suppose what internal space on top of that, it should be flat. But is it possible to imagine a figure that will in same being rolled up and flat?

To understand this chacha, let's turn to a two-dimensional analogy. Let an example flat space will be ordinary paper sheet. To unfortunately him there is four the edges, a our a task in volume and consists, to from these edges get rid of. casket opens simply. If you roll the sheet into a tube, only two open faces remain. on the opposite ends formed cylinder. By connecting them joint in joint, we we get a figure resembling a bagel or donut. In geometry, such a figure is called torus. Topology - chapter mathematics, studying most general properties geometric figures - claims what at similar kind continuous transformations that we just done, the surface of the sheet of paper remains flat. And although on the first sight at the torus With paper sheet general at all a little, surface bagel - good example final flat space.

Among other things, the donut model gives a good idea of why additional dimensions of space are hidden from us, not observable in principle. At The torus has two diameters. The first diameter is "large", this is the diameter of the circle, which was formed when we turned a straight paper tube into a closed ring. Diameter room two a lot of less - this is, simply saying thickness tubes. Suppose what big diameter It has astronomical dimensions and is 1030cm, in then time how small diameter does not exceed 10-30cm. Then a hypothetical creature of average height, dwelling on the surfaces torus, will to seem like his the world is one-dimensional.

So, we have answered the question of how the interior space can be simultaneously flat and folded. Remains figure out With privileged the position of three large dimensions. Why only three spatial coordinates our peace swollen how on the yeast, a all others stayed shriveled little ones? In other words, why is the Grand Universe three-dimensional and not two-dimensional? or, shall we say four-dimensional?

Let's remember scenario chaotic inflation Andrew Linde, about which walked speech in previous chapter. To visually demonstrate uneven character inflation in different domains (or areas) universe, we then took advantage analogy with a plastic film, broken into a kind of checkerboard, each of which has the Planck size. These fields behave purely individually. In some inflation ends relatively quickly, in others it continues indefinitely, and still others collapse instantly, barely having time to be born. plastic film can be stretched as you like and in any direction, so as a result we get kit elementary cells different size and forms.

The same is the case with the predominance of three dimensions. One checkerboard in our model can be stretched evenly, and after the end of inflation, it will still remains a plane, only larger. And the other can be turned into the thinnest a thread whose length will exceed its width by an astronomical number of times. Ant, crawling along such a thread, will quite rightly consider that his world has only one spatial measurement - length, because the width applied practically in zero.

AT scenarios chaotic inflation our real physical Universe is a small part of a huge whole - the Mega- or Metaverse (in English literature the term multiverse is used by analogy with universe - "universe"). "There, beyond the river," far beyond the event horizon, there are other worlds with a different number of spatial measurements unfolded to cosmological scales. They have nothing to do with our universe, and even time in those other universes need not correlate with ours. Speaking cloth language strict science, we With you we live inside one causally connected area, once and for all fenced off from other domains ruled by ball at all other physical laws. Us simply lucky: if would number "big" measurements was two or four, to be interested in the structure of the universe, more likely Total, became would simply no one. By happy chance we were born in world, allowing the formation of complex structures; more precisely, it is only in such a world that we could be born, because universes with other values of fundamental constants are worked out not about us - remember about jewelry at the construction site initial parameters.

So close interest to theories Kalutsy - Klein and problem rolled up (compactified, as physicists say) measurements are by no means a whim and not a game of beads, since they are most directly related to string models. At temperature of about 10^{32} degrees, all four interactions - electromagnetic, weak, strong and gravitational - must merge into a universal single superpower. However traditional performance about elementary particles how about point objects not allows consistent tie general theory relativity With quantum mechanics. In 1984, physicists Michael Green of Queen Mary College London and John Schwartz from Californian technological Institute showed what problem easily is solved if the world of elementary particles is depicted not in the form of tiny spheres, but in the form extended objects, a kind of threads, or strings (strings), having elastic properties. True, for the first time they started talking about strings back in the late 60s of the last century, but until 1984 strings models remained candid exotic, not more how brilliant game mind.

If a stretch elastic rubber tape, tension inside her sharp will increase. But one has only to let it go, as the elastic forces will instantly return to the tape original form. Something similar happens with the string. As the temperature drops the tension of the string increases, and when the temperature drops noticeably below 10^{32} degrees, it immediately shrinks into a point. That is why the elementary particles that we observed today behave like point objects. However, in reality, the fundamentals universe lie invisible strings, elastic character which implies what they can vibrate like a guitar string. Thus, all elementary particles are quarks, electrons, protons - essence not what other how vibration these tiny strings, the longitudinal size of which is comparable to the Planck length (10^{-33}cm). The shorter the length waves, topics above her energy. BUT because the energy is equivalent to mass (remember famous Einstein formula $E = mc^2$), then we can easily compare the length waves and her energy With mass. That's why fluctuations strings With various frequency may interpreted as different particles. This unorthodox approach is amazing. that makes it possible to consider all elementary particles in the form of one and the same fundamental object - strings. Another attractive feature of string theories is that the interaction between particles is elegantly and naturally explained falling apart strings on parts or connection individual her fragments.

So, all famous us bricks universe can liken sounds arising from the vibrations of a guitar string, and then the universe will turn into a grandiose a symphony majestically emerging from the invisible Nothing. Needless to say, impressive and exciting spirit painting, leading on the memory the first opus Friedrich Nietzsche -
"Birth tragedy from spirit music." AT brackets note what strings theories more often called superstring theory because they have so-called supersymmetry, unifying particles With whole back (for example, photons) and half-whole back (for example, electrons) in single diagram, but we are in these physical jungle not climb.

The trouble is that the mentioned strings stubbornly refuse to sound in the space of three dimensions, and therefore superstring theory is valid at least in the ten-dimensional world (one time and nine spatial dimensions, with six of them curled up and hidden from the observer due to their microscopic size). As you know, the guitar string fit fluctuations only With some quite certain length waves, because what her ends hard fixed. Superstrings too hesitate not anyhow how, because the limited internal space - six hidden measurements closed on the myself. That's why length waves, permitted on the string, determined structure and dimensions of the interior space. Thus, the structure of the internal space plays leading role in features those strength, which we we observe.

circumstantial analysis strings theories (a them on the today's day proposed quite a lot) is not included in our tasks. We only note that, say, the so-called M-theory, which is the direct successor of various superstring theories and is very popular, imposes additional restrictions on the number spatial measurements. This model, built in 1995 by a professor at Princeton University Edward Whitten, devoid of obvious contradictions, apparently only in space 11 or 26 measurements. However, superstring theory has not only ardent admirers, but also no less fierce opponents, who rightly believe that the idea of our multidimensionality The universe must be accounted for by the serious difficulties of this model. Other her a significant drawback (despite the mass of advantages that they never tire of reminding apologists strings) is impossibility experimental checks (on extreme least in foreseeable future). And in general, frankly speaking, superstring theory is still very and is far from complete. True, many physicists do not lose hope that the string an approach early or late will allow build universal theory, which received call theory Total (in English - theory of everything, abbreviated TOE).

imaginary time Stephen Hawking

*No, not the moon, but a light dial
shines me and how it's my fault,
What faint stars do I feel milky? And
Batyushkov's arrogance is disgusting to
me; Which the hour? his asked here
BUT he answered curious: eternity.*

Osip Almond stem

So, string theories of various kinds claim to be a consistent unification quantum mechanics With general theory relativity and like would allow forever and ever get rid of annoying singularities with their uncomfortable infinities. However, we already had a chance to make sure that, despite the indisputable advantages, the theory of superstrings frankly slips, when trying hard flatten all multicolor peace to one and only fundamental entities - elastic one-dimensional string, lost in multidimensional space. Therefore, many experts are trying to find other options for bypassing the singularity, offering their own evolution scenarios universe, not related With puzzling geometry. Gol on the fiction cunning, and alternative structures proposed great lots of, but model outstanding British theoretical physicist Stephen Hawking, who prioritizes the concept of the imaginary time, deserves, in my opinion, a separate discussion. However, before talking about imaginary time necessary properly figure out with time ordinary.

Time is generally a mysterious category. From time immemorial, people have been interested in the question of what it is. represents near - an immutable law governing the movement of the worlds, or some psychological kunstuk, through whom our consciousness arranges flow incoming from outside sensations?

More recently, a little over 100 years ago, even great scientists had no doubt in absoluteness time. dials, scattered on boundless universe, everywhere showed the same hour. The universe was drawn in the form of an empty dimensionless box, where majestically circling planets and stars, obeying relentless laws heavenly mechanics. Synchronize clocks scattered at random around the back streets of this giant bubble, it was easier steamed turnips - spit and grind.

Theory relativity not left from these naive stone representations on the stone and today we we know what world arranged a lot of more difficult. Idea absolute time (how, however, and absolute space) ordered for a long time live. Watch two observers, located in different reference systems do not have to match. Today space and time not consider isolated, a unite in universal four-dimensional the "space-time" continuum, which, in turn, is inseparable from material bodies, filling the universe. If a some miraculous way extract from universe all filling his things, all matter before latest particle, then space and time will automatically cease to exist. However, intelligent people understood this. and before. to me already had to quote Christian philosopher Blessed Augustine, who said that the world was not created in time, but together with time. AT his "Confessions" he wrote:

...

If a same before sky and land not It was time then why to ask, what You did then. When not there was no time was and then.

Australian theoretical physicist Floor Davis in book "O time" collected rich a collection of aphorisms about the nature of this mysterious substance - sometimes ernicheskie, sometimes frankly ridiculous, and sometimes exceptionally profound. Let's quote some from them.

...

Mystic XVI century Angel Silesius: "Time created you by ourselves, this is watch in your head. AT that moment when you stop think time too collapse dead."

...

The ancient Roman poet Titus Lucretius Kar: "And in the same way, time cannot exist self on yourself but only from movements of things we get we sensation time. Nobody, we admit, does not feel time in itself, but knows about time only by the movement of everything other things."

...

Bishop James Usher (1611 year): "Start time fell out in night the day before 23 October 4004 before the new era.

...

Inscription on the wall toilet: "Time - this is simply one trouble per another".

...

Christian author Agathon: "Even God not maybe change past".

...

George wheeler, physicist: "Time - this is way, which nature not gives everything take place straightaway".

...

withrow, too physicist: "Time - this is intermediary between possible and realized."

Davis could also remember the unsurpassed Lewis Carroll. When Alice has a cup tea said, what loves not bad spend time, insane Hat indignantly shouted:
"Look what you want! If you knew old man Time like I know him, you wouldn't even know about it. stuttered. Its not spend! Not on the attacked like that!"

Finally, Ostap Bender, whom Davis probably does not know: "The time that we have this is money, which at us No".

However, jokes aside. Time, if you look closely at it, turns out to be in eminently unintelligible concept. Why do we remember the past, but do not remember future? Why, one wonders, in space you can move in any desired direction, along everyone three his axes or coordinates then how's the time fundamentally one-dimensionally and always flowing from of the past in future? Exists even concept "arrows time", and received allocate three her constituents - thermodynamic, cosmological and psychological arrow. Surprisingly, they are all aimed at one side. To the man on the street, these questions may seem idle and without sense, because our unconditional involvement in the flow of events seems to him to be something taken for granted. Meanwhile, the mystery of the "arrow of time" is one of the most difficult, and final response on the question, why time flowing in one quite certain direction, not managed find bye more nobody.

The matter is aggravated by the fact that the laws of science do not distinguish the past from the future. If a talk more strictly, they not are changing in result violations So called CPT symmetries. The letter C denotes the replacement of a particle by an antiparticle, the letter P denotes a mirror reflection, when left and right are reversed, and the letter T - a change in direction movements all particles on the reverse, then there is turn time back. Other words the physical processes that take place in our Universe, will not change one iota if reverse parameters C, P and T. On the other hand, if the laws of science are so are indifferent even to the triple combination of the operations C, P, and T, we are justified in assuming that in the same way, they should not change when performing a single operation T. However absolutely obviously, what between movement forward and back in time lies distance huge size. porcelain a cup, having fallen co table on the stone floor, is bound to shatter, and no one has yet seen the reverse sequence of events when the fragments are brought together, and the whole cup again jumps up on the table. Similar behavior dictated second start thermodynamics, which says that in any closed system, disorder (or entropy, which is the same) always increases with time. In a sense, this best of all worlds is subject to famous law murphy, according to to whom sandwich always falls oil painting way down. Reader gravitating to scientific severity, can suggest several other formulation of this comic law: from two equally probable events always occurs most unpleasant.

So, the law of non-decreasing entropy, or the increase in disorder over time, underlies the thermodynamic arrow. Cosmological arrow reflects expansion universe, a psychological defines our subjective sensation time. BUT because the she is given thermodynamic arrow and subordinate her, we remember events in the same order as entropy grows. That is why we remember the past, and not future.

Initially, to moment Big explosion, Universe stayed in highly ordered state, but on measure Togo how era changed era, a world generated structure after structure in the form of stars, planets and galaxies, entropy steadily grew. At first glance, we are faced with some contradiction, since evolution Universe generally and evolution organic peace in particular (not speaking already about becoming reason on the planet Earth), seemed would, not agree With increase mess. After all, life developed from the simple to the complex and eventually produced light amazing and perfect mechanism - homeostat second kind, what is human brain. Hardly anyone would argue that a person is much more complicated bacteria. Tem not less this is contradiction imaginary, for local orderliness certainly accompanied growth entropy. Stephen hawking illustrated this is circumstance very clearly. He writes to "Brief stories time":

...

If you memorize every word from this book, then your memory will receive about two million units information and order in your head rise about on the two million units. But bye you read this book, on extreme measure one thousand calories orderly energy, which you got in form food, turned in disordered energy that you have transferred into the air around you in the form of heat for convection and perspiration. The disorder in the universe will increase by about twenty million million million million units, what in ten million million million times the indicated increase in order in your brain - and this happen only in volume case, if you remember all from my book.

So the way our subjective sensation time - his relentless psychological arrow - given arrow thermodynamic, and second Start thermodynamics in such a formulation of the question becomes almost trivial. Mess grows with time because we measure time in the direction in which it grows mess. Logics quite impeccable. Remains only figure out, why and the cosmological and thermodynamic arrows also point in the same direction. casket opens simply. If a Universe will be expand enough for a long time, then to to that time when the expansion is replaced by contraction, all the stars will burn out safely, and the particles fall apart on the elementary bricks. Others words Universe will be in extremely disordered condition. But for evolution organic peace and existence reasonable life necessary how we remember strong thermodynamic arrow, becausethat all living things consume food, which acts as a carrier of an ordered form energy. Life translates it into a disordered form, turning the energy of food into heat. Thus, at the stage of compression, the existence of complex structures is impossible, because the world is different extreme disorder and not contains necessary construction material. In addition, during the compression phase, the temperature and pressure will rise steadily, so that any organic inevitably will die in flames of the world fire.

Justice for the sake of should Mark, what some scientists consider newborn universe how extremely disordered structure. Let's say famous Belgian physicist Russian origin Ilya Comely believes what story Universe With moment Big explosion there is not what other how process evolutionary complication of some "primary atom", which was her elemental chaotically homogeneous state. And observable and absolutely indisputable the processes of thermodynamic degradation of our world are purely local in nature and neither in slightest degree not affect on the destiny Universe. By Prigogine processes self-organization will continue unlimited for a long time, bye in end ends not will triumph above forces universal decay. However majority physicists With Prigogine strongly disagree and regard the initial state of the Universe as an example of a highly ordered structure. One way or another, but the question of the arrow of time is still very far from final resolution. And you, the reader, if you want to understand problem more thoroughly, I recommend fascinating book Stephen Hawking "Brief story time."

Let's get back to option bypass singularity, proposed Hawking. running a little ahead, I note that his script is crammed with puzzling mathematics and therefore very not easy for popular presentation. Even at specialists, dog eaten on the various models of the universe, sometimes give up when they try to understand constructions British theorist. For example, famous domestic physicist I. BUT. Smorodinsky frankly writes, what will pass more quite a few time bye alluring andpromising Hawking's idea will become any understandable.

The standard model of the universe, burdened with a singularity, can be graphically portray in form inverted cone, delivered on the point. vertical axis on thesuch a diagram will denote time, and two mutually perpendicular horizontals - the space of our world. The top of the cone corresponds to the point "zero", the moment of birth universe "out of nothing". It is easy to see that the scale factor, that is, the size of the universe, too was equal to in then time zero. FROM flow time diameter circles continuously grows as the universe expands. Thus, our inverted cone can be introduce how kit slices various diameter, each from which corresponds some very specific moment in time. The further back in time (top to bottomvertical axes), topics less the size universe, bye in top cones (then there is in singularities) he finally not will turn in zero. So, before us his kind taper bread loaf, consisting from individual slices of bread.

However, the singularity, as we remember, is not just a dimensionless point, but vanishingly small volume, lying in areas Planck lengths ($10-33$ centimeters). Let me remind you what quantum fluctuations, which we easily neglect in "big" world, become very significant at scales of the order of $10-33$ centimeters. Planck length of a beam of light crosses per $10-43$ seconds, Consequently, we Can consider this value how a kind of "quantum of time". Thus, mother nature herself put up on our slingshot paths that prohibit accurate measurements. The order of things laid down in the original structure of the world, turns out to be stronger than our desires. But as soon as the space and time cannot be physically measured below the Planck limit, it is unclear whethersimilar quantities though any physical meaning. If a on the top cones it is meaningless to talk about space, then exactly the same is true for time at start began.

Let's get back to our cone chart, where time moving vertically up, a space unfolds horizontally and described circle With mobile diameter. At Planck's limit, there, where run amok quantum fluctuationsspace and time finally lose all physical meaning, and we no longer have right to say that time creeps up and space stretches horizontally. Time insuch a model completely loses its inherent specificity, and it is no longer possible to distinguish it from other spatial dimensions. In other words, when the size of the universe was less Planck's limit, time in our habitual submission not existed. AT dark, how known all cats sulfur, that's why time in areas Planck lengths becomes fully equivalent spatial measurements, forming together With them four-dimensional sphere. And only when Universe stepped over Planckian limit and became irresistibly grow, quantum fluctuations lost his fundamental meaning, a space and time found various properties.

Hawking suggested that the universe at the beginning of the beginning was as simple as it is possible. But what could be simpler than a sphere? Therefore, we decisively and irrevocably discard the vertex in our inverted cone model and replace it with the bottom edge round bowl or sphere. From the point of view of the British theorist, space-time is lower Planckian length recalls sphere, and Universe, so the way not It has no start, in volume sense, that she not It has edges or borders.

For visibility let's turn to two-dimensional analogies. look on the ordinary school the globe, this imperfect model earthly ball, and Imagine yourself on the moment that its South Pole will be the birth point of the Universe. Just like from of a stone thrown into the water, circles diverge on the mirror of the pond, so and from the conditional point, timed in this case to the South Pole of our small ball, the Universe starts confidently expand. At this distance from circles to circles, drawn along the meridian will reflect the growth of the universe over time. Clear, what each subsequent a circle will be more previous, bye swelling peace not will reach equator. FROM this moment circles start once per at once decrease in diameter and eventually finally come to naught at the point of the North Pole. And although in such a model, the Universe automatically acquires zero dimensions at both poles, about clumsy singularities can safely forget. Because the all points on the surfaces spheres absolutely equal and nothing not different friend from friend, at growing universe in the scenario of Stephen Hawking is missing a certain special point (that is, singularity) in which all standard physical laws would be violated. Reaching maximum on the equator, latitudinal circles start straightaway same diminish bye not converge to a point at the North Pole. And although the size of the universe is zero at the poles, these points (quite, however, conditional) will singular only on definition, how Southern and Northern poles on the surfaces earthly ball. Laws physics will be carried out in them With such same laid-back ease, how they performed on the Southern and Northern poles planets Earth.

To unfortunately so graceful and smooth description stories our peace requires introductions imaginary time. And although expression "imaginary time" sounds, to be maybe, somewhat wild, it is nevertheless a rigorous scientific concept. If multiply any ordinary (or real) number on itself, we will get an intelligible result positive number. (Let's say two times two equals four, and exactly the same the same is obtained by multiplying -2 by -2.) However, there is a special class of numbers (their received call imaginary), which at multiplication on the myself give negative size. For example, the imaginary unit (usually denoted by the letter "i") when multiplied by minus 1 gives itself. Sometimes it is described as the square root of minus one. In such limiting conditional world with the category of time in the area of Planck lengths occur amazing metamorphoses: it forever loses its original properties duration and starts remind extended spatial measurements. AT at dusk, objects lose their face, becoming similar to each other up to complete indistinguishability.

And only as the scale factor grows, Stephen Hawking's imaginary time acquires my originality. It how would is born on the smooth place, imperceptibly sailing out from space and shaking off yourself an unnecessary tinsel its length.

At first glance, Hawking's scenario may seem like a frivolous mathematical fun. His puzzling calculations are reminiscent of the famous parable of the mad tailor, who sews all sorts of clothes, not in the least caring who they might fit fit. The warehouse of finished products has long been littered with a variety of rags that maybe come up to whom whatever - octopus, centaur, unicorn or cuttlefish. He professes a thoroughly functional approach: each of the clothes is perfect itself by oneself, but a real subject who could pull one or different outlandish outfit, on the horizon not seen. insane tailor related With mathematician installation on internal consistency: the suit can be anything ridiculous, but if it is tailored in full accordance with the rules of cutting and sewing, then already most has the right to exist. Who can really benefit from this crooked hoodie, no role plays.

They say what once outstanding Russian mathematician P. L. Chebyshev set out to read to the Parisians lecture about mathematical theories construction clothes. The quorum was great. The best cutters came to listen to the world celebrity, fashion designers and legislators Maud. holding back breath and pricked up feathers, workers needles uncovered their notebooks and notes books. Chebyshev started from afar.

– Gentlemen, he said, let us assume for simplicity that the human body has the form ball.

Rest the words he agreed in empty Hall.

jokes jokes, but mathematics also not bastard sewn. To the theory Stephen Hawking specialists relate quite Seriously, although and understand her With fifth on the tenth. ingenious the Briton believes that in reality the world lives according to the laws of imaginary time a So called real time - Total only fiction, appearance, a one-day butterfly fluttering over the surface of heavy and imperturbable motionless water. According to his deep conviction, the real time counted by our chronometers, in surroundings Planck quantities is being transformed in time imaginary, and then uncomfortable singularities may to be easily crossed out from stories our Universe. Real time, with which we are accustomed to deal, turns out to be a psychological twist, comfortable notion, phantom invention our psyche, a on the bottom universe the thing-in-itself, imaginary time, rests indifferently. However, let's give the floor to ourselves. Hawking.

...

Perhaps one should conclude that the so-called imaginary time - it's on in fact, time is real, and what we call real time is simply the fruit of our imagination. In real time, the Universe has a beginning and an end corresponding to singularities that form the boundary of space-time and in which the laws of science. AT imaginary same time no no singularities, neither borders. So what to be maybe, exactly then, what we call imaginary time, on the himself deed more fundamentally, a then, what we call time real - this is some subjective performance, arising at us at attempts describe, which we see the universe. After all
‹...› a scientific theory is simply a mathematical model that we have built to describe observations: it exists only in our heads. So it doesn't make sense to ask what is real - "real" time or "imaginary" time? It's only important which from them more suitable for descriptions.

Summing up hell under reasoning daring british, remains Mark, what not the borderless, smooth Hawking universe, for all its charm, It has on extreme measure one significant flaw: practically complete absence evidence-based experimental base. However, there is no reason to believe that the the foreseeable future, such evidence will appear. However, this is not the worst sin, because the lion share other cosmological models too not lends itself experimental verification. Theory chaotic inflation Andrew Linde is, perhaps a happy exception in this series, for it agrees remarkably with the last achievements observational astronomy.

On the other hand, the notion that space and time form a smooth closed surface provides a wealth of food for thought regarding the role of God in life Universe. Philosophical potential this models difficult overestimate. Barely whether not all cosmological scripts, postulating birth peace "from nothing", let implicitly and With big creak, but still allowed Existence Creator.

Stephen hawking writes:

...

If the universe is really completely closed and has neither boundaries nor edges, then it should have neither beginning nor end: it simply is, and that's it! Does it then remain place for the Creator?

Another scenario for the origin of our universe was proposed by an American physicist Lee Smolin. In his opinion, new worlds can be born inside black holes. About black holes, these coal bags universe, where matter fails without return, detailed in chapter Star Panopticon, so I won't repeat myself. Let me just remind you that the black hole stage is a natural stage in the evolution of very massive stars. When star burns his nuclear fuel, internal pressure already not maybe counteract the forces of gravity, and the celestial body collapses inward. Such catastrophic contraction is called gravitational collapse. However, not only stars or other massive objects can be the source of black holes; inflation theory predicts, what on the early stages evolution universe, in phase inflation, must were in multitude form primary black holes.

The gravitational forces inside the event horizon of a black hole are so strong that the collapse continues until the density of matter becomes infinitely large. Samo it goes without saying that the volume occupied by the compressible matter will then vanish. Inside black holes sits already acquaintance us singularity - dimensionless dot With infinitely large density and curvature of space-time. black space holes are a road to nowhere, a bottomless and black as wax failure, from which you can't break out neither one particle. Even light becomes eternal her a prisoner for power gravity per horizon events transcends everything conceivable limits.

However, the theory of relativity, as is known, is not takes into account quantum effects, and therefore, it works very badly on scales smaller than the Planck length. Meanwhile role quantum fluctuations below the Planck limit, when themselves concepts of time and spaces finally lose their physical meaning, it becomes decisive. Same most fair and for curvature space-time. Other words we entitled assume that there is no singularity with its tiresome infinities inside black holes No, a such options, how density substances and curvature spacetime, must to be limited some critical value. But if gravitational collapse in areas Planck lengths coming off on the No, then quite probably, what space inside black holes maybe undergo impetuous bloat. Remember inflation, which increased the volume of a newborn by orders of magnitude Universe? The theory claims that something similar could happen to a black hole, when collapse natural way fizzle out.

However, we are immediately confronted with an unresolvable paradox. If space inside the black hole begins to swell by leaps and bounds, then its volume must multiply grow in a very short period of time. By the end of inflation, it it is easy to exceed the size of the observable part of the universe, if the inflation continued enough for a long time.

But on the other hand, a black hole is a true thing in itself, from which nothing, even light, not can get out out. Any extension, how would great it neither It was, must necessarily be limited by the internal volume of the black hole, its gravitational radius. And since the event horizon of a black hole is no match for dimensions complete universe, then absolutely unclear, what way so grandiose volume maybe fit in inside tiny wicks.

To deal with this paradox, we again have to resort to the two-dimensional analogies. Imagine yourself children's air ball, on surfaces whom crawling a flathead is a tiny intelligent being who is not familiar with the third dimension. In our model, the surface of the balloon corresponds to the three-dimensional space of the universe. From the point of view of a flat person, a black hole in his world is just a small area surface, a pitch-black spot where he can't access. Having traveled around spots, flatfish without labor find out what black hole It has quite final sizes. Now imagine what gravitational collapse inside flat black holes ended a long time ago, and it is going through a phase of rapid inflation. Wherein the rubber of a balloon inside the event horizon is not stretched into a two-dimensional world flat, a swells in direction, directly perpendicular surfaces ball. Places for such inflation more how enough, that's why subsidiary Universe, born before our eyes, can easily surpass the mother in volume. However, for flatfish this process will remain secret per family seals, for his imperfectonly two dimensions of his boring world are available to vision. He won't see at all nothing new: the same inexpressive spot will obtrusively loom before him, although in reality it already for a long time unfolded in huge universe.

Something similar can happen in our real three-dimensional universe. By graduation collapse space inside black holes starts irresistibly expand, and a few moments later, according to the galactic clock, the newly minted world solemnly emerges from non-existence, giving birth along the way their own space and time. To unfortunately us not destined to be witnesses this exciting spectacle, like just as a flat man, with all his desire, cannot penetrate into the third dimension. the universe inside which black hole passed in inflationary mode, we entitled name maternal (or parental), and the "young woman" that budded from her - the daughter, or infant. Both these universes will connected peculiar umbilical cord tube spacetime, diameter which comparable, on all visibility, With Planckian length.

However, the umbilical cord may also break, since black holes, albeit slowly, but evaporate losing mass per check quantum fluctuations near their borders. Horizon events steadily cringes how shagreen leather, and how only he will become less Planck limit, the black hole will effectively shrink to zero, and any communication between related universes stop. Mother and baby heal independent life. Truth, some physics claim what quantum effects suspend evaporation black holes near Planck's limit, but fundamental values this is circumstance not It has. burst connecting them umbilical cord or remained in intact and safety does not play any role: both Universes are still isolated from each other and lead yourself as quite independent creatures.

AT further subsidiary Universe maybe go on footsteps his mothers. When inflation will stop, and the energy of the inflaton field will drop to the minimum values, happen Big explosion, and daughter will pass in mode standard Hubble extensions. After graduation inflation fluctuations density in newborn Universe will become cosmologically significant, which will lead to the formation of primary black holes. Some of them will also begin to swell in turn, so that the light will appear already the third generation of worlds. In a sense, these new worlds will already grandchildren of the original parent universe, who, as time passes, also nearly for sure will give offspring.

Thus, we come to a fundamentally different picture of the universe, which could be called the global universe. The global universe is difficult ensemble worlds and recalls grape bunch. Some grape-universes tied between yourself umbilical cord through black holes, which not managed bye evaporate, a other a long time ago live isolated, but inside most offspring continue to be born primordial black holes, which time after at once they give a start in life to more and more new generations of universes. In other words, global Universe capable continuously self-reproduce. Such tireless budding maybe continue unlimited for a long time, that's why can to tell, what the global universe has no beginning in time. If the vegetative cycle does not stopneither on the instant and works how watch, then global Universe will be live forever.

Of course, each individual grape (or domain, which is local the universe reliably isolated from their brothers) maybe have own unique set physical parameters. Them related only commonality origin, so to speak, the voice of blood. Some worlds, not having time to swell properly, immediately same start collapse, collapsing in point, a other will, vice versa, unrestrained inflate because inflation is growing exponentially. Among all conceivable universes there must be at least one where the inflationary expansion will stop in time, giving rise to density fluctuations, which will subsequently give rise to complex structures - galaxies and stars. By happy chance, we we live how once exactly in such universe. If a would fundamental constants have had other values, this book never not was would written.

breeding budding universes by no means not are twins. Genealogical kinship not It has to them thin structure smooth account no relations. World constants are not Mosaic commandments written on tablets. God is notspoke from the burning bush with the plenipotentiaries of the chosen people, and therefore fundamental constants may accept arbitrary values. Number spatial measurements, deployed before cosmological scale, too not must be limited to the number "three" and may vary significantly in individual local bubbles. Even time inside budding grapes maybe throw out amazing knees and flow like a god on the soul positive

We can't look inside black holes, because everything that is done for horizon events, under impenetrable lid spheres Schwarzschild, represents an absolute terra incognita. But if Lee Smolin is right, and our Metagalaxy, in other words observable universe, hatched at one time from a primordial black hole, we have neitherhow not comparable possibility study her offal from within, quite simply exploring structure around us peace.

Us remains reply on the the only one sacramental question. By most According to modest estimates, the mass of our Universe is approximately 1022 solar masses. But if The universe is so hefty, how can this abundance of matter fit into tiny volume of a primordial black hole? In fact, no paradox here even not smells. Let's remember what creation peace "from nothing" suggests equilibrium between negative energy gravity and positive energy substances. BUT because the the negative gravitational energy exactly compensates for the positive energy, related With mass, giving as a result, zero, the mass of the child the universe can be very big. Newborn baby maybe without special labor outgrow his parent.

One could put an end to this, but more recently, the English physicist Barbour introduced on the court most venerable public sensational book under name "End time." In it, he set out to prove that no time exists in nature, andthe sequence of events that we habitually arrange along the time axis is not what other how inertia our thinking, not having With reality nothing general.

This is perhaps the most radical and extravagant hypothesis about the nature of time, and my story about various cosmological models was would incomplete, if would I not paid concepts brisk Englishman although would several lines. Theory Barbour detail reviewed by Raphael Nudelman in the fascinating article "The Newest Guide to time", published in two rooms magazine "Knowledge - strength" per 2002 year, and you, reader, without labor you can With her familiarize. I I will retell her in short.

Barbour occupy not real physical objects, but the relationship between them. If a we let's take three points and connect them direct lines, then we get triangle certain kind. it and will be Barburovsky "ratio" which describes three point system. If at the next moment of time the position of points in space change, then the triangle will take a different form. This new "ratio" will have already other characteristics.

Now let's denote the length of each of the sides of our triangle by some number. We construct a space with three coordinate axes and on each axis we set aside one of these numbers. At us succeed the only one dot in space, which, self yourself of course will be reflect not real position initial dots, a Total only relation between three objects. We call such a conditional space (with a physical space, it has nothing in common) configuration space, or K-space. All subsequent states of the system of three objects throughout its history will be be described totality dots, certain way pierced on K-space.

Similar operation can do With each from real physical particle, filling the universe and then all they will take due them place in configuration space. Barbour calls his fictional space Platonia in honour great Greek philosopher which the, how known insisted on the real existence of common concepts (universals). According to Plato, the material world is pale copy magnificent peace ideas his flawed likeness.

If a would world obeyed laws classical mechanics newton, each subsequent his condition definitely flowed would from the previous one. Such painting universe was called deterministic and was in full swing right up to the beginning the last century. The outstanding French astronomer and mathematician Pierre Laplace time even undertook to calculate the future of the universe, if he had at his disposal accurate coordinates of all elementary particles. In Newton's world, an observer outside Platonia, could would specify subsequence all Barbour points and connect them trajectory, in result what at each points appeared would "story" co their own past and the future.

To unfortunately we we live in probabilistic world, which the controlled quantum laws. Principle uncertainty imposes fundamental ban on the simultaneous definition coordinates and speed elementary particles: how more precisely one parameter is measured, the less accurately another can be calculated. Therefore, the points on map of Platonia, reflecting the position of particles in the configuration space of Barbour, should replace probabilities. But then picture At once will lose sharpness: instead fixed points we we'll see light haze, trembling haze above red-hot asphalt. Connecting them with a rigid trajectory (writing a "history") will be decisively impossible.

But why do we still perceive time, if in reality it does not exist? Barbour argues that "the impression of change arises here only because in our the brain collects several portions of information about various positions (or states) the same object." In his opinion, the rejection of the category of time allows not only get rid of singularities with their heap of infinities, but also once and for all deal with his clumsy arrows. In all other cosmological scenarios, time flowing from of the past in future, because what Universe had Start. But in Platonia Barbour
"moment zero" missing on definition, because the time from her taken out. If a in Platonia for three points available some special configuration Alpha, where all particles are in one place, then then same most fair and for Universe in in general.
"Big" Platonia must also have its own Alpha configuration, a special highlighted point, when all particles the universe are in one place.

Barbour writes:

...

The landscape of Timelessness unfolds like a flower to all other points that present yourself universal configuration most different sizes and difficulties. Maybe, the form Platonia is what promotes reinforced downstream probabilistic
"foam" in side those configurations, which contain "reminders" his general origin from point Alpha.

One word, time, With points vision Barbour - this is phantom, disembodied ghost, product of our imperfect psyche. We perceive it as a stream that has quite a certain direction, only because we ourselves are an integral part of this peace, his unconditional offspring. True Universe deprived time his brings there our stupid consciousness, which, by hook or by crook, strives to see in the unknown painfully familiar, putting on a tailcoat pair on an octopus, and therefore describes the world is purely approximately.

What can be said about this? The vast majority of physicists estimate Barbour's ideas are very skeptical, rightly believing them to be empty mathematical fun, set brilliant scholastic paradoxes. Not sorry places and let's quote famous Australian theorist Paula Davis.

...

Barbour, roughly speaking, claims that time does not really exist. I'm ready agree that space and time are not the ultimate realities. It is possible that underlying them reality represents yourself some "PRE-Space-Time", from elements of which our observable space-time is constructed, just as the substance we observe is built of microparticles, which, in turn, can turn out to be built from PRE-particles, from more more fundamental building blocks matter - superstrings - or something in this kind. Like particles substances space-time too may be derivative concept.

...

And yet, at a sufficiently large level, on the scale of the macro- and mega-world, this most spacetime, which us familiar. From him it is forbidden simply get off With with the help of mathematics ... At one time, before the advent of the theories of relativity and gravity, in it was fashionable in certain circles to say that the time It's just a human fruit consciousness, derivative from our Feel flow events, what is it somehow way associated with the ability of the brain perceive events only in some "temporal sequences." There's no denying that time is a flow, but it's not purely human invention or category of consciousness. For a physicist, time and space together with matter are this is part toy structures, With which is born herself Universe or, more precisely, from which created Universe. Talk, what time is not exists, simply meaningless.

Here So, short and clear. Ivashka will go walk, a Vitka will be at home sit, how spoke one worker-peasant mother, annoyed behavior his senior son.

It remains for us to spit with the so-called anthropic principle, after which we let's move on to more burning questions. About amazing alignment fundamental constants, jewelry fit initial parameters universe us happened to be mentioned more than once, and you, dear reader, must remember what is wrong the devil is terrible, how they paint him. Cosmological models postulating multiplicity worlds (for example, theory chaotic inflation Andrew Linde or unrestrained budding global Universe Lee Smolina), allow throw away on the landfill hackneyed hypothesis of the Creator, because they consistently and consistently decide question about thin at the construction site world constants. However, not we will run in forward.

The term "anthropic principle" was first proposed by a professor at Cambridge University of Brandon Carter, one of the greatest astrophysicists of our time. However, shrewd people paid Attention on the amazing alignment fundamental constants long before Carter. So, in the early 1950s the famous English astrophysicist Fred Hoyle wondered how carbon and oxygen in stellar bowels. To him managed notice curious numerical the ratio between the total energy of three alpha particles (or equivalently, nuclei helium) and the energy level of the carbon nucleus. So that when three alpha particles merge, carbon, this magnitude must make up 7.7 megaelectronvolt. Subsequently this the quantum effect was discovered experimentally. And a little later the great Paul Dirac caught more one amazing match between sizes observable universe and force gravity in her, although these quantities no way not connected friend With friend.

Not less interesting and that fact, what density our peace very close to critical density. If a would magnitude ? was several less critical, scattered matter, which is in a very rarefied state, simply would not have time to get together in masses, necessary for formation stars. FROM another hand, if ? more
?cr, then, on the contrary, condensation will proceed at an accelerated pace, and life in the Universe (or, speaking more strictly, complex structures) simply will not have time to arise. And already all the more, there would not be enough time for the evolution of the organic world, which on Earth, as known lasted several billion years.

If a increase in 100 once numerical meaning gravity permanent, then in so many same once will be reduced time life Sun. Clear, what fifty million years clearly not enough to on the planets solar systems arose biosphere. At others values of the electromagnetic interaction constant, the proton will lose its stability - the fundamental brick of the universe, and if, in addition, we "correct" the constants a little strong and weak interactions, appearance Universe will change before unrecognizability.

Relations masses proton, neutron and electron between yourself too have determining value both for the modern structure of the Universe, and for the appearance in it life. Let's say that the mass of a neutron exceeds the mass of a proton by a negligible amount (about 10-3m?). If we just double this value, the atoms of the chemical elements lose stability. Similarly, an increase in the mass of an electron by only a factor of three will lead to decay nuclei atoms hydrogen - most widespread element in Universe.

Dimension surrounding us space too gives rich food for reflections. Three spatial dimensions ensure stable circulation of bodies friend near friend: or body stable moving on ellipse (in private case, on circles), or flies away in infinity on parabolic or hyperbolic trajectories. But in the four-dimensional world, periodic motion in a closed orbit impossible: planet or will fall on the central light, or immediately fly away in infinity. This means that in the world of four spatial dimensions, exist sustainable planetary systems, the movement of electrons around atomic nuclei and etc. All matter crumbles into dust. And in worlds with less than three dimensions, atoms lose ability radiate in continuous spectrum, because the electrons not may there make the necessary for this orbital transitions.

List the thinnest adjustments fundamental constants continuously growing and today has already reached a truly frightening magnitude. Slowly think about what someone wise and prudent deliberately polished the universe so that it could grow up human. AT our disposal available three option response on the this question.

Option one. The laws of nature are created by a higher mind. Theoretically similar situation quite possible that's why not we will become reject her With threshold. AT end ends we create in earthly laboratories artificial nutrient media for growing beneficial micro-organisms, and who knows what further advances biotechnology will make through a couple of thousand years. However, it is not entirely clear how this hypothetical the higher mind managed to survive in the flames of the universal fire when our world emerged from non-existence, and where he was and what he was doing when the world did not yet exist. On the other hand, overmind - it is also supermind in Africa, and our business is calf - pissed off your legs and stop. However serious scientists begin to wince when it comes to divine providence. FROM help intervention supernatural forces can without any labor explain any phenomenon, but then the science orders for a long time live. natural science an approach, in difference from faith, inclined confess principle Occam: not should multiply number entities in excess of need. That's why let's leave option room one theologians and theologians. higher strength are listed according to them department.

Option two. If Theory of Everything (TOE for short) will ever be built, it is likely that the numerical values of the fundamental constants will receive a natural and reasonable explanation. When scientists understand what is mass, charge, spin and other purl essences of the universe, perhaps it will be possible to answer to the question why they take exactly these and not other values. Then anthropic principle can will be take off With agenda day. M-theory about which told in previous chapters, today claims on the much, but before finish line bye more long away.

Option third, most kind our heart. If a universe not exhausted observable part universe, if they in multitude born from quantum fluctuations space-time foam (on Linde) or from primary black holes (according to Smolin), then our Universe ceases to be unique and the only one in his kind. Fundamental constants may accept in these countless worlds any arbitrary values, and life and intelligence arise only in those universes where conditions are right for them. True, it may seem to some that nature on the rarity wasteful: pile up sort of break through worlds, to in some from them a spark of reason ignited. Einstein said that God does not play dice. Meanwhile there is nothing to be surprised here, because nature is a blind constructor, and wastefulness is her immanent feature. Of the millions of eggs, only a few thousand survive, and the trees every year they scatter the seeds in abundance, so that some of them grow. On the question "Why is our universe the way we see it?" the answer follows: "If Universe was another, us would here not It was!" it and there is wording anthropic principle.

In fact, the anthropic principle exists in two formulations - weak and strong. The weak anthropic principle insists that intelligent life arises only there and when and where the conditions are right for it. Let's say modern cosmology claims that the universe began about 14 billion years ago and will continue to exist long enough. Why do we live relatively close to the moment of her birth? The casket opens simply: 10 billion years ago, second-generation stars with chemical the composition necessary for the appearance of complex structures was not yet, and after a few tens of billions of years, they will all burn out without a trace, and intelligent life of our type will become impossible.

Strong anthropic principle says what laws nature and options fundamental constants are such that allow the emergence of intelligent life. Other In other words, the world is imprisoned for man. Frankly, both versions of the anthropic principle claim practically one and then same, but after all his strong hypostasis considerably gives back teleology. It turns out, what all gigantic machinery universe conceived solely for you and me. It is not easy to reconcile with such a formulation. Besides, not bad would contribute one significant clarification.

When we we say what at others parameters fundamental constants in Universe impossible complex structures and life, should would add: life in forms known to us. But even protein life on planet Earth has a huge adaptive potential. Enough recall about So called "black smokers" – hot geysers on the bottom oceans. They are are hereby hotbed life, although the temperature of the water near the smoker reaches 300 degrees Celsius at a pressure of several hundreds atmospheres. What already here talk about hypothetical extraterrestrial organisms whose exchange substances maybe build up on the fundamentally different chemical basis.

Incidentally, the natural and climatic conditions of our planet are changing very rapidly. wide range, which does not interfere with earthly organisms feel great and pole, and on the equator. Temperature optimum - a business taste. Nile crocodile would have had a hard time in the Arctic Circle, and polar bears, walruses and seals would hardly liked would tropics. If a would white bear was able reason logically, he would certainly come to the conclusion that wise nature took special care to ensure that to him, white bear It was OK. Not bring Lord live in sultry desert, where afternoon with fire you will not find tasty and healthy seals, and the sun burns mercilessly. Is it a matter of relatives Penates: cool some water, fresh breeze and semi-annual polar night…

Summing up this chapter, we emphasize once again: the idea of a plurality of universes natural way allows problem jewelry settings fundamental constants, so the cumbersome and clumsy hypothesis of God can be with a clear conscience safely send off to junk.

Ring around sun

On the distant star Venus The sun is fiery and golden, On the Venus, ah on Venus At trees blue leaves…

Nicholas Gumilyov

If the universe was exhausted by galaxies, stars and other black holes, we could would boldly put here point. However in world there are more and planets - compact non-luminous body, circulating around stars, and on the one from such heavenly tel we live we With you. Word "planet" in translation With Greek means
"wandering". The ancient Greeks, several centuries before the birth of Christ, noticed that in extensive family motionless stars there is their fidgets, drawing on the firmament confused curves. antique astronomers knew five wandering stars - Mercury, Venus, Mars, Jupiter and Saturn. Together with the Moon and the Sun they made up cosmos of the ancient world, and the sphere of fixed stars crowned this slender architectural ensemble like domes. Earth, by itself, was center universe.

Subsequently, the magnificent five was replenished with three more eternal wanderers - Uranus Neptune and Pluto. This trinity it is forbidden make out unarmed eye, therefore, it was discovered relatively late - after the invention of the telescope. Uranus discovered in 1781 by the English astronomer William Herschel, Neptune in 1846 - French Urban joseph Le Verrier, a Pluto - American Clyde William Tombo in 1930s Truth, Pluto, for a number of reasons, is today denied the right to be called a planet and placed in special category dwarf planets or transneptunian objects.

AT our days even pupils junior classes know what around what spinning.

The central place in the solar system belongs to our daylight, and the planets apply Around him along elongated circles - ellipses.

Correctly draw orbits planets managed long away not straightaway. Myself creator heliocentric systems Polish astronomer Nicholas Copernicus thought what orbits planets are regular circles. And only after more than 100 years another famous astronomer, German Johann Kepler, managed show, what the only geometric figure consistent with observational data is an ellipse, and the Sun situated in one from his tricks.

Relatively sizes sun too existed various opinions. Most desperate ancient Greek minds admitted that it could be the size of Athens, and one sage, daring suppose what Sun already no way not less Peloponnesian peninsula, was exiled With disgrace. Of course true dimensions sun several more. And although it occupies a modest place in the stellar nomenclature, being considered ordinary yellow dwarf class G, its size is very impressive. The diameter of the Sun is about 1.4 million kilometers (diameter of the Earth for comparison - a little over 12 thousand kilometers), and in German concluded 999/1000 all masses solar systems. Average distance from Earth before sun - 149 million kilometers. This value received call astronomical unit (a. e.), and she is serves for measurements interplanetary distances. The Sun is one of the 200 billion stars that inhabit our Galaxy (the Milky Way), and is located along with its nine planets in the outskirts of the galactic spirals, in 26 thousand light years from her center.

Let's take a closer look at the structure of the solar system. Except four planets terrestrial groups (Mercury, Venus, Earth and Mars), four gas giants (Jupiter, Saturn, Uranus, Neptune) and in many all more enigmatic Pluto in compound solar systems are included So called small planets, generators belt asteroids between orbits of Mars and Jupiter, as well as comets and meteors arriving from its distant outskirts. There, beyond the orbits of Neptune and Pluto, a belt extends for dozens of astronomical units. Kuiper - a collection of dwarf planets and rock and ice fragments of various shapes and sizes. Still further away lies a huge spherical cloud of protoplanetary bodies, named in honor of the Dutch astronomer by the Oort cloud. From there, long-term comets. Finally, at majority planets solar systems there are natural satellites (except Mercury and Venus). Jupiter currently has over 60 satellites, Saturn has 56, Uranus has 27, Neptune has 13, and Pluto has 3. Mars Total two satellite (Phobos and Deimos, what in translation With Greek means "fear" and "horror"), and our old Earth managed to acquire only one thing - the Moon. But but nearest neighbor Earth looks very impressively on the background others satellites, yielding in size only to the three largest satellites of Jupiter (Ho, Ganymede, Callisto) and satellite Saturn Titan.

Among the ancient Romans, Mercury (aka the Greek Hermes) was considered the god of trade, and since the alpha and omega of commercial transactions has always been deception sell, they say), then in combination this cunning god patronized rogues and scammers.

How and befitting glib and efficient shopkeeper space Mercury green agile: it runs around the Sun in just 88 days, and its year, therefore, in four superfluous times shorter earthly. Distance before Mercury from sun is changing in wide within - from 46 before 70 million kilometers, making up in average 58 million kilometers. It is easy to see that the orbit of Mercury resembles a strongly elongated ellipse, which differs markedly from the almost circular orbits of all other planets in the solar systems. Ellipticity orbits heavenly body received to express through her eccentricity
– the ratio of the major and minor semiaxes of the orbit. In the case of Mercury, this value is 0.2, while the eccentricity of the earth's orbit is more than 10 times less (about 0.017). Except Togo, orbit Mercury perceptibly tilted to ecliptic - plane terrestrial orbits. Corner

tilt is 7 degrees. For these two parameters - the degree of eccentricity and the angle inclination to the ecliptic - only Pluto managed to surpass Mercury (0.25 and 17 degrees respectively).

Due to its proximity to the Sun, Mercury receives six times as much sunlight per unit area than the earth. At perihelion, the point of minimum distance from the Sun, temperature his illuminated surfaces is 430°C a in aphelion - point maximum removal - drops to 290 ° C. Temperature on the night side of the planet falls before minus 170°C. Since the average density of Mercury almost like that same, how at Earth, it must have an iron core, which, according to calculations, takes up almost half volume planets.

From the surface of the Earth, Mercury is quite difficult to observe through a telescope (in medium latitudes he not bad visible only in summer months), that's why compose for real reliable maps of the planet and to clarify its physical characteristics turned out to be possible after Togo, how neighborhood nearest to sun planets visited space probe
"Mariner-10". Mercury is small and very hot, it is inferior to the Earth in diameter by almost three times, and by volume - 14 times. The diameter of Mercury is 4880 kilometers, and the mass is 5.5% of the mass of the Earth. The force of gravity on its surface is three times less than that of the earth, and a man of average height would weigh about 25 kilograms there. Among the planets of the solar systems smaller than Mercury only distant Pluto. Mercury has an extremely rarefied helium atmosphere created by the solar wind and containing negligible amount hydrogen, traces argon and not she. Her pressure at the surface planets in 500 billion times less than the air pressure on Earth at sea level. Probe "Mariner-10" also revealed that Mercury has a very weak dipole magnetic field (100 times weaker earthly).

On the throughout long time astronomers thought what Mercury, how and Moon to earth, always converted to sun one hemisphere, then there is revolves around axes synchronous with the movement around the Sun. However, in the mid-1960s, with using radar research, it was found that the period of rotation of the hot planet of the solar system is about 59 days, therefore, Mercury makes a complete rotation around its axis in two-thirds of its year. Logically, solar gravity must was a long time ago slow down his axial rotation, but stake soon this did not happen, a tempting hypothesis arose that Mercury had once rotated around Venus and only relatively recently was rejected by a more massive celestial body. In any case, mathematical modeling of its orbit does not exclude the possibility what in distant past he was satellite Venus.

Named after the ancient Roman goddess of love and beauty (among the Greeks - Aphrodite) Venus - nearest our neighbor among big planets (least distance from Earth
– only 39 million kilometers) and the brightest star in the night sky after the moon. She is shines in 13 once brighter Sirius to whom belongs honorary first place the brightest stars. The brilliance of Venus is so great that with a certain skill it can sometimes be seen even during the day, against the blue sky. This is because the second planet from the sun wrapped in a thick atmospheric coat, 100 times more powerful than the Earth's atmosphere. Gas cover Venus, permeated several layers clouds, great reflects sunlight.

Honour discoveries Venusian atmosphere belongs our compatriot Michael Vasilyevich Lomonosov. watching in 1761 year passage Venus on solar disk, he wrote: "A bump appeared on the edge of the Sun, which is all the more the closer Venus got to the performance. Soon this pimple was lost, and Venus suddenly turned out to be without edge ... "Lomonosov concluded that" the planet Venus is surrounded noble air atmosphere... What is drenched near our ball earthly."

Venus located nearly in one and a half times nearer to the sun how Earth (108 and 149 million kilometers respectively), a because receives from bounty our luminaries in two With half the heat. In terms of size, Venus and Earth are almost twin sisters: the diameter of Venus is only slightly inferior to the diameter of the Earth and is 12,104 kilometers (0.95 of the earth's diameter, which is equal to 12,756 kilometers), and its mass is equal to 81% of the mass Earth. Full turnover around sun Venus commits per 225 earthly days, a here periodits rotation around the axis is somewhat larger - 243 days. No other planet in the solar system does not rotate around its axis so leisurely, Venus is the undisputed record holder on parts most slow daily rotation. In addition it committed inside out, in side, opposite her orbital movement, what actually not unique property Venus. Let's say Uranus and Pluto too spinning in reverse side, but they do this lying almost on their side, while the axis Venus almost perpendicular to the plane orbits. Thus, she is the only one planets, which "really" revolves vice versa. Figure out how should in features of the daily rotation of Venus was succeeded relatively recently - in the early 60s years of the last century, when radar methods began to be widely used, which made it possible look in under her dense cloud cover.

Before flights to Venus first space probes many science fiction writers imagined our nearest neighbor as a sort of tropical paradise, sultry and stuffy peace, covered impassable jungle. In wet dusk boundless selva vile creatures were hiding, busy devouring their own kind. Unlike decrepit dying Mars Venus drawn some scientists junior sister Earth as it was in distant geological epochs, many millions of years ago. Others insisted that there was no land on Venus at all, and the entire surface planets occupies a continuous vast ocean.

Reality turned out where more prosaic and more unexpected. It turned out that the atmosphere
The "white-faced beauty" (as the astronomers of ancient China called Venus) is 96.5% from carbon dioxide and almost 3.5% from nitrogen. And for the share of all other gases - oxygen, water vapor, sulfur oxide and dioxide, argon, neon, helium and krypton - do not have to more than 0.1%. True, it should be borne in mind that since the Venusian atmosphere is 100 times more powerful than the earth, it contains about five times more nitrogen than in the Earth's atmosphere. On the surfaces planets, under monstrous cloudy bedspread, reigns unprecedented, deafening heat in 460–470 degrees on Celsius. At such temperature are melting some metals. Even the sunlit side of Mercury is a little cooler. And although powerful cloudy layer thick in several dozens kilometers reflects 77% falling on the him sunny Sveta, oversaturated dioxide carbon atmosphere creates the strongest greenhouse effect on the surface of Venus, due to which the temperature and reaches such high values. For the same reason, it is surprisingly stable and does not depends on the latitude of the area. Only in the highlands it is a little cooler - on several dozens degrees.

Cloudy layer, containing droplets concentrated sulfuric acid, extends to a height of 70 kilometers, and in the uppermost layers of the atmosphere there are also hydrochloric and hydrofluoric acids. Cloudy layer revolves how single whole, but much faster than the planet itself, making a complete revolution in 4-5 days. Therefore, at the heights about 60 kilometers hurricane-force winds constantly blow at a speed of 100 meters per second (360 km/h). But near the surface of the planet, the wind speed drops to several meters per second, but since the atmosphere of Venus is 50 times denser than Earth's and only 14 times inferior in density water, then even wind force one meter in give me a sec - very serious trial. The pressure of the atmosphere on the surface of Venus is 90 times that of the earth (90 and 1 bar, respectively), and a pressure of 119 bar was recorded at the bottom of the Diana Canyon. Even on the highest mountain peaks of the second planet, reaching 11 kilometers in height, pressure is 45 bar, that is, 45 times more than on Earth at sea level. In a word, Venus - this is world sizzling heat, purged through red-hot winds and forever and ever crushed severe carbon dioxide fur coat, few inferior on density water.

Of course, no life in the forms we are accustomed to can survive in the hot inferno. second planet. The white-faced beauty of Chinese astronomers turned out to be the most real hellish fiery.

For a short century of terrestrial astronautics, about thirty automatic stations. The first descent vehicles were designed for maximum pressure of about 7 bar, and therefore quickly collapsed even in the upper layers of the Venusian atmosphere. But it was with their help that it was possible to establish the gas composition of the cloud cover our closest neighbour. Domestic probes Venera-13 and Venera-14, which made in 1982, a soft landing on the surface of the planet, managed to work for about 2 hours in murderous climate Venus. Analysis soil showed what minerals, components bark planets, in many similar earthly basalt, meeting on the bottom oceanic deep water basins. American probe "Magellan" for four years of work in orbit Venus (1990-1994) compiled and transmitted to Earth detailed maps of its surface. Relief second planets complicated and represents yourself extensive hilly plains, crossed by numerous ridges resembling mid-ocean ridges on the earth, and also alpine volcanic plateau origin.

Volcanic activity Venus doubt not calls. On the her surfaces tens of thousands of volcanoes have been discovered, some of them reaching 100 kilometers in across. It is possible that individual volcanoes continue to erupt to this day, but their the number is relatively small. Completely unique landforms have also been identified in form very fat and slowly spreading lava flows - So called pancake volcanoes. But there are very few meteorite craters on Venus - about 900, that is, not more two on the million square kilometers. For comparisons: on the mars on the such same area there are nearly a hundred and fifty craters, a on the moon - near four hundred. Apparently, this is due to the fact that in the recent past (about 500 million years back) its surface has undergone a kind of renewal: ancient rocks with traces meteorite bombardment were filled with young lava. An additional argument in the advantage of just such a scenario is the absence of manifestations of plate tectonics on Venus, typical for Earth or Mars.

That's why in last thing time became very popular hypothesis So called
"sudden volcanism", designed to explain unique climatic features Venus. According to this hypothesis, the absence of continental drift led to the fact that slowly accumulating underground heat about half a billion years ago overnight splashed out through tens of thousands of simultaneously emerging volcanoes. In atmosphere planets received monstrous amount carbonic acid, untwisted flywheel greenhouse effect. The result of these processes was the disappearance of water and the rapidpromotion temperature.

It remains to be added that Venus has not been found to have a magnetic field or radiative belts, despite the presence of an iron core with a radius of 3000 kilometers and a powerful mantle from molten breeds, occupying a large part volume planets.

The fourth planet of the terrestrial group received the name of the ancient Roman god of war Mars, which the originally was chthonic deity fertility and wild nature. The Greek word "chthonos" means "earth", and it is customary to call chthonic creatures creatures of the earth's interior, abundantly endowed with its productive power. Valiant Mars became a warrior later and as such was identified with the ancient Greek Ares, patron of insidious and perfidious war for the sake of war, while Athena Pallas personified war honest and fair.

Mars is one and a half times farther from the Sun than the Earth, so the Martian year twice as long as Earth's: its duration is 687 Earth days. Besides, the orbit of Mars has a rather noticeable eccentricity (0.09), so that the distance to fourth planets from sun is changing in tangible within - from 250 million kilometers in aphelion before 207 million kilometers in perihelion (at Earth relevant values are 152 and 147 million kilometers). Average distance between Mars and sun is 227.9 million kilometers.

Features of the Martian orbit lead to the fact that every two years (more precisely, then every 780 days) Earth and Mars are at a minimum distance from each other friend, which ranges from 56 to 101 million kilometers. Similar planetary encounters are called confrontations. If the distance between them becomes less than 60 million kilometers, then they speak of a great confrontation. This event is repeated through every 15–17 years old.

The diameter of Mars is 6800 kilometers, that is, it is almost half the size of the Earth. In terms of mass, it is 10 times inferior to our planet, and in terms of surface area - three and a half. times. A Martian day is slightly longer than an Earth day (24 hours 39 minutes and 23 hours 56 minutes). respectively), and the angle of inclination of the equator to the plane of the orbit is 25 degrees, which only two degrees greater than that of the Earth. However, unlike our planet, the seasons in the northern and southern hemispheres of Mars have different durations, which is explained by conspicuous the elongation of its orbit.

One word, Mars on many parameters very similar on the earth, much more, how any another planet solar systems, that's why he always called increased interest among earthlings. The course of reasoning was extremely simple: if on Earth during time it flourished life, then is it possible to exclude that Mars is inhabited planet? And as soon as he, in all likelihood, is older than the Earth, then there is quite maybe exist highly developed civilization, much ahead of in technical relation to the earth. When, at the end of the 19th century, the Italian astronomer Giovanni Schiaparelli reported what repeatedly saw on the surfaces Mars net long dark lines, binding polar and moderate zones planets, American Percival Lovell immediately suggested them artificial origin. Following per scientists to cause writers joined in, throwing fuel on the fire from the bottom of their hearts. The fascination with Mars grew beyond days a by the hour.

H. G. Wells populated the fourth planet with hideous giant slugs with a tuft of tentacles around a beak-shaped mouth. The product of a completely different evolution, they were embodiment naked reason With cleanly clipped emotional sphere. Arrogantly and contemptuously they looked from the cosmic heights at the stupid swarming earthly life. Our planet was interested in these intelligent cephalopods solely as an inexhaustible food resource, as another outpost on the path of their irresistible expansion. Far ahead of earthlings in technical terms, they easily built a huge interplanetary fleet, and at the turn of the century (Wells' novel The War of the Worlds was written in 1898 year) Martian spaceships fell like peas on the long-suffering Earth. clumsy armies Europeans turned out to be not in forces resist gigantic and invulnerable combat tripods, smashing on the spot all surrounding deadlythermal beam. Cities came in desolation, a iron roads overgrown weedy grass. advancing the end Sveta. Humanity saved accident: martian ruined terrestrial microorganisms harmless to people, because they lived in their homeland practically in sterile conditions, nearly fully having lost immunity, So how more a lot of centuries back exterminated all infectious and parasitic illness. amazing carelessness for highly developed civilization, mastered manned space flying…

Fundamentally other interpretation clashes two worlds proposed Alexei Nikolayevich Tolstoy in the fantastic story "Aelita" (1923). He sends to Mars two enthusiasts - the engineer Los and the Red Army soldier Gusev. After a short interplanetary flight apparatus, built at the expense of the republic (where is the money, Zin?), safely landing brave travelers on the surface of the Red Planet. Mars under the pen Tolstoy irresistibly rolling to sunset. This decrepit, dying world a long time ago

ineptly squandered the heritage of the great past, and now a high culture created by hard work of dozens of generations of Martians, is in deep decline. withered channels, abandoned residents cities, destroyed before grounds gigantic reservoirs
– on the everyone lies seal ruin and desolation.

Along the way, it turns out that with its unprecedented cultural takeoff Martians are obliged natives from the planet Earth: 20 thousand years ago, when the legendary Atlantis, having split into pieces, sunk into the depths of the sea, the ferocious Magatsitls - the supreme caste of the Atlanteans, by fire and planted with a sword civilization around the world - began leave native planet. Through ocean falling water, in smoke and ashes, they flew away in world space in bronze, who had form eggs, space devices. Martian annals Togo time say:

Forty days and forty nights the Sons of Heaven fell on Tuma. The star Talzetl was rising after evening dawn and burned with an unusual light, like an evil eye. Many of the Sons of Heaven fell dead, many were killed on the rocks, but many reached the surface of Tuma and were alive.

The ancestors of the Martians called their home planet Tuma, and the bloody star Talzetl - this is Land in local dialects. The aliens plowed the fields and sowed them with barley, cut through barren martian plains network channels and erected cyclopean the buildings. Together With them came Great Knowledge, recorded colored spots in ancientmanuscripts.

Messengers Soviet Russia caught at all another era. If a take advantage terminology of Lev Nikolaevich Gumilyov, a well-known Russian historian, Martians finally and irrevocably lost passionarity and fell into sheer insanity. Similar condition, when society extremely atomized a vital energy his members fluctuates near the freezing point, it is commonly called the obscuration phase. Shards of high culture decayed in dusty book depositories, a power was usurped bunch cynical oligarchs. The common people lived in poverty. It goes without saying that the hero civil war, retired division commander Gusev, endure such ugliness not could. He looked around, and his soul became wounded by suffering. Combat division commander started a military coup, and at first fortune favored him. But things soon went haywire. overturning loose militia rebels government troops have crossed in resolute offensive, and our heroes had to hastily carry away legs. Turn on fourth planet in compound Russian federation, to unfortunately So and failed.

So pages "Martian chronicles", released from under pen American science fiction writer Ray Bradbury, a very different Mars rises. But in the most poignant In the short stories of this cycle, we see the same thing - a fragile, refined culture that is dying under the boots of unceremonious and uneducated colonists from Earth. These strong and vigorous guys wonderful know With which sides at sandwich oil, a slightest manifestation intelligence causes at them healthy life-affirming laugh. They are fun shooting down toy Martian towns long abandoned by their inhabitants, and weightless porcelain turrets silently crumble into dust. Endangered natives somehow survive mine century in most deaf and inaccessible corners planets, and only rarely-rarely can see impetuous snow-white sailboats martian, cutting the sharp stems of the red sands of the Martian deserts. And at the crossroads mushrooms after rain, grow up ugly cans, open sausage under clumsy signs and heavy trucks purr as they clumsily turn around in the clouds thin orange dust. One word, repeats great American frontier, in result whom perished and melted away without trace unique culture the whole continent.

BUT what is fourth planet in reality? What represents yourself real, a not an imaginary Mars? Until recently, there were no answers to these questions. Scientists fantasized who in what much. Mars is a dead planet, some said. If there was life, then she is perished hundreds million years back, when on Earth walked around antediluvian lizards. Nothing of the kind, others objected. And what do you want to do with an extensive network channels (by the way, up to 50 kilometers wide!), Which connect the polar caps with temperate latitudes of Mars? There is no doubt that these are complex irrigation buildings, redistributive precious martian moisture. Rave gray mares, skeptics fumed. The so-called channels are just natural faults martian crust. And who said that Mars is a harsh and ancient world, enthusiasts asked. Perhaps most of it - oceans bound by an ice shell, and the notorious channels - quite simply cracked ice or vegetation, fed subglacial moisture.

Relative clarity came only with the beginning of the era of astronautics. The first probes reached the fourth planet, registered an extremely rarefied atmosphere, the complete absence of any large reservoirs and numerous traces of intense meteor bombardment. Today, when in the vicinity of Mars (and on its surface in including) visited many automatic stations, we have the right to bring the first preliminary results. And if passage popular film actor Filippova before now since remains unanswered ("Is there life on Mars, is there life on Mars - this is still science unknown") then relatively flowering apple trees can speak out more definitely.

Since Mars receives more than two times less heat from the Sun than the Earth, average annual temperature on its surface is minus 60 degrees Celsius. And although in the summer at the equator the temperature sometimes rises a few degrees higher zero, daily temperature drops are huge and reach several tens of degrees. For example, in the southern hemisphere at the fiftieth parallel, the temperature at the height of autumn does not rises above minus 18 degrees Celsius at noon and drops to minus 63 at night degrees. So significant scope temperature hesitation on the throughout days explained extreme sparsity Martian atmosphere, consisting on the 95% from carbon dioxide gas. On the share nitrogen and argon account for 2.5% and 1.6% respectively, a the oxygen content does not exceed 0.4%. registered on the northern polar cap exclusively low temperatures order minus 138 degrees Celsius. atmospheric The pressure on the surface of Mars is 160 times less than on Earth at sea level. Just on at the bottom of the deepest depressions, it "grows" twice. The Martian atmosphere is extremely dry and nearly fully deprived water vapors. In addition on the mars periodically flare up the strongest storms, lifting in air billions tons dust. Them duration comes before 100 days, a speed wind reaches 70 kilometers in hour.

So the way modern Mars - this is very severe world, and talk about the existence of any complex life forms in such extreme conditions, according to all probability, not account for. FROM another hand, not should forget, what life characterized by extraordinary plasticity and high adaptive potential. We already happened mention about communities organisms fabulous myself feeling near
"black smokers" on the oceanic day, where temperature reaches 250–300 degrees Celsius. Some earthly bacteria may manage without oxygen and survive in acids and alkalis. The solid surface of the Earth and the oceans are only a small part inhabited peace, a deep in bowels our planets flourishes complex ecosystem microorganisms, nearly not communicating With external the world. By opinion some scientists, amount organisms settled under earth, noticeably exceeds number ground inhabitants. controversy many bacteria may in flow long time survive in space, what It was not once proven experimentally. Of course hard ultraviolet light kills them, but a thin protective layer of dust, as a rule, turns out to be quite enough, to significantly increase them resilience.

Therefore, it is absolutely not excluded that in the Martian soil can be found primitive life forms, especially considering the fact that there is water on Mars. The bottom layer of the Red Planet's polar caps, several kilometers thick, is complex from ordinary water ice mixed with dust, and on top they are covered with a thin film frozen carbon dioxide. it So called "dry ice", which the for sure Good to you sign, reader: his wide use in summer heat, to save from premature melting some food products, for example ice cream. Between by the way, seasonal changes in the polar caps are associated precisely with the evaporation of this thin (about 1 meter) of the top layer. In addition, in some areas below the surface of Mars must be located many kilometers thickness eternal permafrost. O availability cryolithosphere testify in in particular some peculiarities buildings geological structures on the surfaces Mars. BUT relatively recently theoretical calculations got reliable experimental the confirmation. American the space probe "Mars Odyssey", launched in April 2001, discovered on the 60th degree south latitude is a vast ocean of subsurface water ice. Moreover, by According to some scientists, in the Martian soil at depths of 100 to 400 meters, water maybe be even in liquid condition: in otherwise case difficult explain the origin of specific furrows on the walls of canyons and craters. True, not really clear, how at creepy Martian cold weather freezing priming on the couple kilometers deep into maybe survive liquid water. FROM another hand, near igneous foci, which on the mars enough, ice maybe melt, passing in liquid phase.

Several discouraging that fact, what reentry devices American "Vikings" committed soft landing on the surface Mars and on the throughout for several years studying the composition of the atmosphere, meteorological conditions and soil, did not find traces organic matter, which could be a product of the vital activity of microorganisms. However quite Maybe, what constructors self-propelled devices quite simply wrong chose direction searches. If a microbes hiding deep in ground, "Vikings" elementary not could them find.

So the way question about Martian life can formulate in three options: one) on the mars never not It was life; 2) on the surfaces planets life No, but she is maybe exist in her bowels; 3) today on the mars life No, but she is existed in the past that's why can find her traces. FROM first option all clear. Relatively second possible various opinions but to confidently reason about subsoil bacteria necessary additional research. BUT here third option represents undisputed interest, because the many scientists convinced what in distant past water on the mars It was in excess. According to some calculations, 4 billion years ago it was even more than on Earth. About this testify grandiose canyons and dried up river riverbed, in found on the surface of Mars . Some of them reach 200 km width at length several thousand kilometers. Even mighty Amazon – the most deep river our planets - looks on the this background enough pale. Where could get rid of water, formed these geological structures, age which evaluated in 3 billion years and more? Between topics planetary scientists not exclude what in that distant era extensive areas northern hemisphere Mars were covered ocean kilometer depth. Dead Martian lakes are also found visibly-invisibly. One of them It was relatively recently identified American geologists. His dimensions may hit most rich imagination: on area it quite comparable With the total territory of Texas and Mexico, and the depth of this monster reached 2 kilometers. So what same after all happened With Mars? Scenarios catastrophes invented great lots of. For example, French astronomer Jacques Laskar believes what corner inclination axes rotation Mars to plane his orbits there is magnitude variable. Today, how known martian axis tilted to ecliptic under angle 25 degrees, then there is Total on the two degrees more, how corner inclination terrestrial axes. By opinion Lascara, 6 million years back this magnitude was 47 degrees. Mars lay practically on the side, and his poles received maximum sunny heat. Polar hats melted away fully, and in atmosphere planets received huge quantities carbon dioxide gas and water vapors. Carbon dioxide provided greenhouse effect, and water couples condensed and fell out on the surface, forming

ocean several kilometers deep. Laskar believes that over the past 10 million years corner inclination Martian axes to plane ecliptic repeatedly changed in very wide within - from 13 before 47 degrees. Cause to that It was powerful the gravitational field of the nearest neighbors of Mars, in the first place - Jupiter. Fourth the planet resembles a children's spinning top or spinning top in state of unstable equilibrium which render impact from the outside. Mars all time "dancing" and poles planets receive then excess, then flaw sunny heat. Today on the mars glacial period. Between by the way, on opinion French astronomer, earthly axis too could would
"jump" back and forth if would not stabilizing influence Moon.

another version catastrophes proposed our compatriot Alexander Portnov, whose article was published in the February issue of the journal "Knowledge is Power" for 2004 year. Mars often called Red planet and in this name No no exaggeration: its surface really has a reddish tint due to the high content in Martian ground So called red-colored sands. Here these completely unusual red sands of Mars, reminiscent of the color of blood, just interested Portnova. A business in volume, what and red color blood, and red color Martian sands explained one and toy same cause - abundance oxide gland. Hemoglobin, imparting blood specific color, contains oxide gland, a his trivalent oxides in form sand and dust cover surface Mars. Portnov writes:

...

American stations handed over intelligence about chemical composition Martian soil and indigenous mountain breeds. These data indicate that the red Martian soil is composed of from oxides and hydroxides gland With impurity glandular clay and sulfates calcium and magnesium. Such a set of minerals is typical for red-colored minerals widely developed on Earth. weathering crusts that occur in a warm climate, an abundance of water and free oxygen atmosphere.

In past geological epochs, when the earth was dominated by warm and humid greenhouse climate, red flowers were common much wider and, probably, covered the surface of almost all continents. The total power of terrestrial red flowers reaches several kilometers, but then same most can see and on the Mars: layer Martian "rust" evaluated in 3–5 kilometers. Between by the way, neither on the one the planet of the solar system, except for the Earth and Mars, such "rust" is not found. At the same time, it is well known that red-colored rocks on Earth could form only after Togo, how in atmosphere appeared free oxygen. But hitch in volume, what practically the whole oxygen terrestrial atmosphere (a his there 21%) It has biogenic origin, that is, formed as a result of biospheric processes. In other words, oxygen - this is product and offspring life. If a destroy all vegetation, free oxygen will evaporate almost instantly. It will reconnect with organic substances will go in in compound carbon dioxide and oxidize iron mountain breeds.

Where did the Martian "rust" come from, if the oxygen content in the atmosphere the fourth planet is completely negligible - no more than 0.4%? Such an amount is clearly not enough for education powerful layer red-colored breeds. Consequently, these the rocks are very ancient and formed when there was a lot of free oxygen. He was removed from the Martian atmosphere and oxidized the iron of rocks, forming the famous red sands. branched river net irrefutably testifies about in abundance water in distant past. Summary: so powerful layer "rust" on the mars could occur only with the combined action of water and free atmospheric oxygen in conditions warm climate. BUT because the oxygen in such quantities must have biogenic origin, to mars once forests roared.

What happened? What killed life on the Red Planet? Portnov believes that The debris of its third moon, Thanatos, collapsed on the surface of Mars. However, about everything order.

Martian red sands have a unique feature - they are magnetic. Often they are called so - the magnetic red sands of Mars. But earthly red flowers, strange the way not magnetized. AT how same is this the case? More once let's listen Portnova:

...

This sharp difference in physical properties explained topics what at the same chemical composition (Fe2O3), the mineral hematite (from Greek "hematos" - blood) With impurity limonite (hydroxide gland), a on the mars the mineral maghemite, a very rare mineral in terrestrial rocks, a red magnetic oxide, predominates iron, having the chemical composition of hematite, but the crystal structure of magnetic mineral magnetite (Fe3O4).

Hematite and limonite are common iron ores, while maghemite is formed occasionally at oxidation magnetite, if persist his primary crystalline structure and magnetic properties. When heated above 200 °C, maghemite transforms into hematite and becomes non-magnetic.

Maghemite was considered a rare mineral on Earth until I discovered that territory Yakutia literally bombarded with huge quantity magnetic oxides gland. These were red-brown sand or patches of various shapes. But the properties of this maghemite were unusual: after calcination he remained magnetic, like his synthetic analogue. I described his how new mineral variety and named
"stable maghemite". arose questions: why he is different on properties from
"usual" maghemite, why his So a lot of in Yakutia but No among numerous red flowers equatorial zones Earth?

It remains to explain where stable maghemite came from, and even in such quantities. Portnov writes that it is easily formed when limonite weathering crusts are calcined, which in Yakutia very a lot of. Consequently, need search source high temperature. At first, scientists sinned on forest fires, but this did not explain even counting nothing: forests are burning everywhere, including at the equator, and magnetic iron oxide is there either not at all, or negligible. The solution came, as often happens, with an unexpected sides.

A giant meteorite crater was discovered in the basin of the Siberian river Popigay near 130 kilometers in across, age whom, on opinion specialists, is
35 million years. grandiose catastrophe happened on the turn two geological periods of the Cenozoic era - the Eocene and Oligocene, when the flora and fauna of the Earth underwent significant changes. In particular, the boundary of these eras is marked by the divergence of a single trunk of primates and the appearance of the first anthropoid monkeys. It is likely that one of the reasons that reshaped the face of our planet was a meteorite attack from space. Presumably, the Popigai asteroid reached 8-10 kilometers in diameter and flew from speed near thirty kilometers in give me a sec. He struck atmosphere through, a released at hit energy was so great, what instantly melted several thousand cubic kilometers mountain breeds, mixing together basalts, granites and sedimentary deposits. Within a radius of several thousand kilometers, everything burned to the ground, evaporated water lakes and rivers, a surface planets on the significant throughout fried like a bone in fire.

BUT now remember what directly per orbit Mars situated belt asteroids - huge Roy miniature planets and debris wrong forms, applying around sun between orbits Mars and Jupiter. The most big from small planets - Ceres, open more in 1801 year, It has diameter about 1000 kilometers, butthe vast majority of celestial bodies in the asteroid belt are much smaller - from hundreds meters to several kilometers. Signs of an intense meteorite impact have been found on Mars. bombing; some only gigantic craters, each from which more Popigaisky, there are more than a hundred on its surface. Thus, we are entitled suppose what magnetic red flowers Mars obliged their origin the strongest calcination his soil in result asteroid hit. sparse the atmosphere of the fourth planet also receives a natural explanation, since gases at high temperatures turn in plasma and disappear in space. BUT oxygen, detectable today on the mars in insignificant quantities, can boldly name relic: these are the wretched remnants of the oxygen that was once generated by the destroyed life.

Mars has two tiny satellites - Phobos and Deimos ("fear" and "horror" in translated from Greek), which revolve around the mother planet at very low orbits. Them origin finally not installed. AT his time famous domestic astrophysicist AND. FROM. Shklovsky even expressed hypothesis what Phobos maybe have an artificial origin, but subsequently his hypothesis was not confirmed. According to most scientists, the satellites of Mars are captured by him from the asteroid belt. They are present yourself heavenly body wrong forms With nearly circular orbits. Phobos resembles a potato 26 kilometers long and 18 kilometers wide. Dimensions of Deimos less - 16 and ten kilometers respectively. Deimos draws around Mars on the a distance of about 23 thousand kilometers, but Phobos creeps very low: it is separated from planets a little less 6 thousand kilometers. Period his appeals very small - per alone martian day he has time thrice go around Mars. Phobos fast approaching to maternal the planet and quite Maybe, what he enough soon (on astronomical standards, of course) will cross So called limit Rosha, then there is some quite certain critical distance (own for everyone heavenly body), on the which gravitational strength tear apart the satellite on the parts.

On the mars limit Rocha passes in 5 thousand kilometers from surfaces planets, therefore, Phobos was a little short of an inglorious but noisy death. Estimated specialists, tragedy happen about through 40 million years and will be have catastrophic consequences. When the debris of the satellite crashes into Mars, its surface warm up before highest temperatures, a leftovers atmosphere in form plasma fly away in world space.

Portnov writes:

...

How we see titles for satellites chosen very successfully: Mars located under Fear with Horror to boot. I think Mars had at least one other moon for which the best name is Thanatos, death. Thanatos was in a lower orbit, than Phobos. He was inhibited by the dense Martian atmosphere, passed through the limit Roche, and its fragments destroyed all life on Mars. Fragments of this terrible asteroid attacks - pieces of the Martian crust - flew to the Earth. Curiously, the craters on Mars form linearly elongated zones and follow friend per friend, how traces submachine gun queues. Maybe, So reflected directions "main strokes" falling friend per friend debris Thanatos.

What can to tell on this about? Version Portnova, undoubtedly, deserves attention because what great explains various inconsistencies in recent geological past Red planets. FROM one hand, dry canyons and prehistoric river valleys, washed relic waters, a With another - dead a lunar landscape that leaves no chance for geologists. When the wreckage of the shattered satellite burned all alive on the surfaces Mars happened magnetization red-colored rocks, and the remnants of the Martian atmosphere turned into hot plasma and scattered in interplanetary space. From cosmic heights descended deadly cold, and per few millions years Mars turned in lifeless desert.

Between by the way, our planet too knew not the best time and not got tired shy away from extremes in extreme. On the throughout recent two million years cruel glaciation With enviable regularity changed warm interglacials. About 10 thousand years ago, in the so-called Holocene maximum, glaciers finally melted away and average annual temperature stubbornly climbed up. Per relatively a shortover time, it has grown very thoroughly, exceeding modern values by 3-5 degrees. AT at that time, all climatic zones were shifted 800 - 1000 kilometers to the north, and latitude contemporary Murmansk noisy oak forests. Desert Sahara was blooming savannah, on the expanses of which boundless herds of ungulates plucked grass, and in the muddy crocodiles and hippopotamuses splashed in the warm pools. But does anyone today this remember? Affairs for a long time past days legends of antiquity deep…

Deserves attention story Alexandra Portnova about flown before Earth fragments of the Martian crust after the fall of Thanatos. Meteorites come from Mars known several dozen, which in itself leads to certain reflections. Today their Martian origin is practically beyond doubt, since the isotopic the composition of the rare gases of these celestial bodies is identical to the composition of the atmosphere of Mars. But the meteorite ALH84001 weighing near 2 kilograms, found in Antarctica in 1984 year, called real sensation. Careful study of the find showed that the mentioned meteorite experienced a strong impact about 16 million years ago, and hit the Earth relatively recently (13 thousand years ago). Everything would be fine, but the study of his inner structures using a scanning electron microscope made it possible to identify in the body heavenly guest very specific details, reminiscent of fossils microorganisms. By character chemical deposits, inside which
"mothballed" bacteria, scientists came to conclusion what them age is 3.6 billion years, that is, it undoubtedly refers to the moment the meteorite was in the Martianrocks. True, experts are confused by the fact that hypothetical Martian bacteria in 100 - 1000 once inferior in sizes them earthly analogues. Microbiologists shake shoulders: in so small volume not will be able fit in intracellular organelles,necessary for their vital activity.

Dimensions "Martian" bacteria quite comparable With earthly viruses, but recent not have cellular structures and not may exist on one's own. FROM on the other hand, to what extent can microbiologists be trusted when it comes to laws stranger evolution? One word, question remains open: to present time in disposal terrestrial science available the only one witness extraterrestrial life, very, however, unreliable.

Fifth planet solar systems on law wears name supreme god from ancient Roman pantheon. Olympian Jupiter, he is the Greek Zeus the Thunderer, severe, but fair mister: to him nothing not costs shy away deadly perun on lousy non-hearing, whoever he was - a man or some other creature of God. To blind one Jupiter, it would take 318 Earths - exactly that many times it surpasses the Earth in mass. And although he is more than twice as weighty as all the others planets of the solar system, taken together, it takes at least 1047 Jupiters to fashion one and only Sun. Diameter Jupiter surpasses terrestrial in eleven once andis almost 143 thousand kilometers. As befits a patriarch of a planetary family, he floats across the sky with dignity befitting his dignity, imposingly and unhurriedly, in accompanied by an honorary escort of his 63 companions, making a full circle around sun per 12 without small years. reigning persons With Olympus hurry nowhere at them ahead eternity.

Jupiter leads list gas giants, which strikingly different from planets terrestrial groups. Firstly, they very great and massive: on the them share account for 99.5% of the mass of the entire planetary family. Secondly, they are composed mainly of hydrogen and helium, therefore, the average density of the substance of the giant planets approaches the density of water - from 0.7 g/cm3 of Saturn to 1.6 g/cm3 of Neptune. The average density of the terrestrial planets is much above and fluctuates from 5.5 g/cm3y Earth before 3.9 g/cm3y Mars. Thirdly, they deprived distinct verge, separating atmosphere and surface planets: them powerful gas shell smoothly passes in ocean liquid molecular hydrogen. Finally, all the giant planets are ringed, but if everyone has heard about the famous rings of Saturn, then similar education at Neptune, Jupiter and uranium were discovered relatively recently.

Regal Jupiter looks very impressively even on the background their gasbrothers. For example, Saturn, which is not much inferior to it in size, is more than three times lighter Jupiter. Visible surface fifth planets - this is layer solid cloudiness from alternating dark and light belts, painted in different colors and extending from equator to the fortieth parallels of northern and southern latitudes. The diversity of latitudinal zones due to the admixture of various chemical compounds. Perhaps the most famous detail on the surface of Jupiter - the so-called Great Red Spot, an oval formation variable sizes, located in the southern tropical zone. At present it dimensions are 15,000 x 30,000 kilometers, so inside the red spot you can labor to lay side by side two globes. Astronomers observe this mysterious structure on the over 300 years.

Some scientists considered red spot solid and enough easy body, floating in upper layers atmosphere, but this extravagant version not found confirmation. According to modern concepts, the Great Red Spot is free migrating atmospheric vortex of the anticyclonic type, however, the origin of this vortex and the reasons for its amazing stability, planetologists cannot say anything certain.

Despite its heftiness, Jupiter rotates very quickly around its axis. Full rotation is completed in just 9 hours 50 minutes, so the duration of the Jupiter days not exceeds ten hours. BUT because the planet represents yourself non-solid body, speed axial rotation differs depending from latitude, so equatorial the zones rotate faster than the polar ones. There are no seasons on Jupiter because the plane of its equator practically lies in the plane of the orbit (the angle of inclination is only 3 degrees). As already mentioned, the main components of Jupiter that make up the body planets are hydrogen and helium in a ratio of 80 and 20%, respectively (by mass). At In this study using space probes showed that the upper layer of cloudiness, in all probability, composed of pinnate ammonia clouds, and below is the mixture hydrogen, methane and frozen crystals ammonia. Per check convective processes in atmosphere of Jupiter, a system of stable zonal currents is formed in the form of strong winds blowing in the same direction. Their speed is very significant and ranges from 50 to 150 meters per second. Jupiter has a powerful magnetic field, according to the strength on order superior magnetic field Earth. planet surround extended radiation belts, a plume magnetosphere Jupiter can fix even per orbit Saturn.

Jupiter is located five times further from the Sun than the Earth, at a distance of about 800 million kilometers, that's why temperature external cloudy cover gigantic the planet does not rise above minus 130 degrees Celsius. However, thermal radiation his bowels twice exceeds inflow sunny heat, what He speaks about complex processes, ongoing in depths planets. FROM depth pressure and temperature swiftly are growing reaching very big quantities. AT 1995 year neighborhood Jupiter visited American probe "Galileo", the descent module of which managed with the help of a parachute penetrate the atmosphere of the gas giant up to a depth of 156 kilometers, as a result resulting in valuable data on the internal structure of the planet. And the probe itself for the first time in history entered orbit around Jupiter and until 2003 studied the planet and its satellites. I will bring quote from fundamental labor "Astronomy: century XXI", released to 175th anniversary State astronomical Institute them. P. TO. Sternberg.

...

On the basis data, received space probes, and theoretical calculations mathematical models of Jupiter's cloud cover were constructed and ideas about its internal structure. In a somewhat simplified form, Jupiter can be represented as shells with density increasing towards the center of the planet. At the bottom of the atmospherethick 1500 km, density which fast growing With deep, located layer gas-liquid hydrogen thick near 7000 km. On the level 0.9 radius planets, where pressure is 0.7 Mbar (then there is in 700 000 once more earthly. - *L. Sh.)*, a temperature is about 6500 K, hydrogen passes into a liquid-molecular state, and after 8000 km - into a liquid metallic state. Along with hydrogen and helium, the composition of the layers includes a small amount of heavy elements. Inner core with a diameter of 25,000 km - metal silicate, including also water, ammonia and methane. Temperature in center is 23 000 K, a pressure - fifty Mbar. similar structure It has and Saturn.

It's clear: Jupiter - this is world, So different from our what It was would too recklessly With threshold reject possibility existence unusual forms life in bowels of a huge planet. Jupiter's atmosphere contains oxygen, nitrogen and carbon and content oxygen, on some estimates, maybe in 5 - ten once exceed sunny. And although search water give the most contradictory results, question about the presence of water vapor in the atmosphere of the fifth planet has not been finally resolved. In every case, the presence of short-lived cumulus clouds in the vicinity of the Great Red Spot makes about much to think.

No less interesting are the large satellites of Jupiter, which are commonly called Galilean, in honor of the Italian physicist and astronomer who discovered them at the beginning of the 17th century Galileo Galilei. There are four of them - Io, Europa, Ganymede and Callisto, and Ganymede is the most big satellite in solar system; he surpasses on sizes even Mercury. However, at present, the attention of most scientists is attracted by the second ofGalilean satellites - Europe as a possible candidate for the role of the cradle of protozoa life forms. The fact is that the surface of this small planet (its diameter is slightly less lunar) is covered with a powerful ice crust of a hundred-kilometer thickness, and under it lazily rollswaves a solid ocean of liquid water, the depth of which can reach 50 kilometers. The subglacial ocean is a kind of mantle of Europe, and it is quite likely that that the water in it is warm, because it is heated by the heat coming from the bowels of the planet. So the way second satellite Jupiter - the only thing, Besides earth, heavenly body solar systems, not testing lack of life-giving moisture.

Medium density Europe approaching to density planets terrestrial groups and is about 3 g/cm3. Consequently, 80% of its mass falls on silicate rocks, composing the heated core, and 20% - on water ice (liquid water-ice mantle plus ice bark). Ice shell planets covered thick network cracks and faults, whatspeaks of active tectonic processes occurring in the bowels of Europe. Large cracks stretch on the thousands kilometers, a them width fluctuates from twenty before 200 kilometers. It is possible that in the warm subshell ocean of the second satellite of Jupiter may exist protozoa forms life. Some scientists believe what most favorable terms must take shape not in oceanic depths, a in areas tectonic faults on the surface of the planet. The fact is that due to the tidal effect Jupiter cracks periodically are narrowing and are expanding. AT last case water rises almost to the very surface, and then the sun begins to penetrate its thickness light, necessary for sustaining life.

Jupiter's other moon, Io, is slightly larger than the Moon and is notable for its active volcanism, which the stimulated tidal impact maternal planets and gravitational perturbations of its closest neighbors - Europa and Ganymede. But almost consists entirely of rocks, and dozens of active volcanoes emit sulfur vapor and sulfur dioxide to a height of hundreds of kilometers at a speed of 1 kilometer per second. That's why at very low average temperatures on the Ho surface (minus 140 degrees on Celsius) there can discover hot spots size from 75 before 250 kilometers the temperature of which reaches 100–300 °C. Jupiter's largest moons are Callisto and Ganymede is half ice. The diameter of Callisto is almost equal to the diameter of Mercury, a Ganymede is superior it in size.

The sixth planet of the solar system, known since ancient times, was named in honour Roman god Saturn whom received identify With Greek Kronos. Saturn had a bad habit of swallowing his newborn children, for, according to the prediction Gaia, he was to be deposed by his own son. Managed to escape the sad fate only junior Zeus-Jupiter instead of whom Rhea wife Saturn slipped husband wrapped in diaper stone. matured, Jupiter committed palace coup, a voracious parent dropped in Tartarus. AT antiquity Kronos-Saturn symbolized inexorable all-devouring time. The personality, to be sure, is unpleasant, although the son with daddy too not especially stood on ceremony. So what poet had complete right write:

And by midnight it rises in the east Dead
Saturn and shines like lead. Truly sinister
and cruel
Your affairs, Creator!

Like Jupiter, Saturn is a huge ball of gas, rapidly rotating around an axis. A day on the surface of Saturn lasts 10 hours and 40 minutes. Although Saturn is not very much inferior to Jupiter in size (its diameter is only 20 s a small thousand kilometers less than the king of the planets, and is 120,500 kilometers), it is more than three times lighter than it, but 95 times more massive than the Earth. This is explained unique low middle density sixth planets: she is less density water and is 0.7 g/cm3 against 1.33 g/cm3y Jupiter then there is nearly twice below. Saturn not able drown even in kerosene.

Saturn is almost one and a half billion kilometers away from the Sun - ten times further Earth, therefore, per unit area, it receives 90 times less solar heat, and its the temperature at the upper cloud boundary does not exceed minus 120 degrees Celsius. However, the thermal radiation of its bowels is twice the energy flux received by it from Sun. Saturn - hydrogen-helium ball, but in difference from Jupiter he contains much more hydrogen on comparison With helium - 94% and 6% respectively (on volume). Orbit this cold giant represents yourself nearly correct circle, a full turn around sun he commits for 29 s half years.

famous rings Saturn first discovered Dutch physicist and astronomer Christian Huygens in the second half of the 17th century, and a quarter of a century later the French astronomer Italian origin J. Cassini managed make out dark slot, dividing the bright flat ring in two. The outer part of this giant necklace, extending nearly on the million kilometers, called ring BUT, a internal - ring B. Subsequently, four more rings were identified - C, D, E and F, and in 1980–1981 American space probes Voyager 1 and Voyager 2 was sent to Earth pictures Saturn and his rings With high resolution. On the these pictures distinctly it is seen, what rings Saturn consist from many thousand individual narrow rings. System rings, girdle sixth the planet - this is myriad stone and icy debris most various quantities and forms.

Saturn is as striped as Jupiter, but due to low temperatures, freezing ammonia vapors with the formation of dense fog, its latitudinal belts are not so clearly visible. A giant oval-shaped atmospheric vortex is found near the north pole size With earth, received title Big brown spots. AT atmosphere Saturn blow strong zonal wind, speed which - from 100 before 500 meters in give me a sec depending on latitude. Like Jupiter, Saturn has a powerful magnetic field, axis which coincides with axis of rotation planets.

Of the 56 moons of Saturn, the most interesting is its largest satellite - Titan. slightly inferior to Ganymede, but superior in size Mercury. Its diameter is 5150 kilometers, but make out details on the surfaces planets not seems possible due to dense atmosphere, pressure which in one and a half times more than on Earth at sea level. Titan's atmosphere is almost entirely nitrogen (98.4%), while methane accounts for only 1.6%. In addition, it contains impurities of propane, ethane, acetylene, argon, helium, carbon monoxide and dioxide, and some other gases. The temperature of the upper atmospheric layers is approaching minus 120 degrees on Celsius then how temperature surfaces planets a lot of below and is minus
179 degrees, what explained peculiar anti-greenhouse effect (thick fog scatters and reflects the sun's rays. Incidentally, if a person by some miracle ended up on Titan, he, in all likelihood, would be able to easily soar in its very dense atmosphere, attaching wings like the Greek Icarus to their hands, since gravity on the surfaces largest moon Saturn in seven times less terrestrial.

Before recent time scientists thought what under cloudy fur coat titan maybe hide ocean kilometer depths from ethane, methane and nitrogen, but data, received automatic station Cassini, visited neighborhood Saturn and become his artificial satellite, forced to reconsider this opinion. At the beginning 2005, Cassini fired off the Huygens probe, which entered Titan's atmosphere and using a parachute, made a soft landing on its surface. It turned out that liquid on the titan very not much: bye managed find only relatively small hydrocarbon lakes near the north pole. After the "titanization" of the Huygens, this the planet became the only satellite in the solar system (not counting, of course, the moon), on the surface which got down space probe. BUT station Cassini continues properly work for orbit Saturn so far since.

Until the second half of the 18th century, no one had ever been born under the sign Uranus, because our ancestors did not know about the existence of this celestial body. seventh planet The solar system was discovered in 1781 by the Englishman William Herschel, for which he was awarded the title of court astronomer with a salary of 200 pounds. Rookie almost immediately dubbed Uranus, which was quite natural: since Saturn is native to Jupiter dad, then another planet should have been called in honour grandfathers.

Uranus spinning around sun on the distance near 3 billion kilometers, making full turnover per 84 of the year co speed nearly 7 kilometers in give me a sec (Earth's orbital speed is 29 kilometers per second). There is nothing surprising in this for the farther the planet is from the sun, the slower it rotates - so says the third Kepler's law. But the axial rotation of Uranus is quite unique: the plane of its equator inclined to the plane of the orbit at an angle of 98 degrees, so that it rotates around the axis almost lying on my side. Therefore, the length of day and night on the seventh planet much exceeds period her axial rotation. Sun, which With surfaces uranium looks bright star, slowly, in flow 21 earthly of the year, rises in sky, a having reached the zenith, another 21 years slowly creeps down until it disappears beyond the horizon. Coming 42 year old night. So this is the case a business on the poles, where duration days and nights is 42 years old. At a latitude of 30 degrees, day and night last for 14 years, and at a latitude of 60 degrees - 28 each. The period of axial rotation of Uranus is equal to an average of 15 hours, significantly changing in depending on latitude.

How and other giant planets, Uranus represents yourself huge gas ball, on the 85% consisting from hydrogen, on the 12 % - from helium and on the 2.3% - from methane. His average the density is only slightly higher than the density of water and is 1.3 g / cm, and the mass is 14.5 timesmore than the mass of the earth. In size, the seventh planet is noticeably inferior to Jupiter and Saturn, however, its diameter (about 51,120 kilometers) is four times the earth's. Uranus is very cold world. The temperature of its surface almost does not change in latitude, but significantly fluctuates depending on the depth - from minus 210 degrees Celsius at the level of the upper cloudiness before minus 170 degrees in subcloud layer. AT difference from others gas giants, Uranus has virtually no internal heat sources. At the seventh planet discovered powerful magnetic field and nine very narrow and dense rings, nearly not reflective sunny Sveta. Before present time in surroundings uranium visited one and only space probe - Voyager 2, rapidly flying past itJanuary 1986.

BUT what maybe to tell the science about giblets lying on the side grandfathers?
AT book
"Astronomy: century XXI" we read:

...

According to the model of the internal structure of Uranus, in the center the temperature of the planet should be lower than that of Jupiter and Saturn, but higher than that of the Earth - about 7200 K, and the pressure near eight million bar. Above big core, consisting from metals, silicates, ice ammonia and methane and occupying about 0.3 of the radius of the planet, there should be a mantle of mixtures of water and ammonia-methane ice. At the level of 0.7 radius from the center begins gas shell from hydrogen and helium.

Uranus is accompanied by 27 satellites, the largest of which, Titania, has a diameter 1580 kilometers. The average daily temperature of the surface of the satellites, 60% of which are ice, extremely low - less than 60 K (minus 213 degrees Celsius). water ice at this temperature turns in solid mineral.

Neptune was discovered in 1846 "at the tip of a pen" by the French astronomer Le Verrier. Having discovered anomalies in the orbital motion of Uranus, he suggested that on the seventh planet solar systems renders influence unknown massive body, and exactly calculated its position in the sky. Guided by Le Verrier's calculations, the German astronomers Halle and D_re without labor found eighth planet which showed up inpoint heavenly spheres, specified perspicacious French. it It was complete triumphclassical mechanics Newton.

It was decided to name the new planet Neptune (aka the Greek Poseidon) in honor of ancient Roman patron of the sea. Storm-ruling Neptune relatives brother Jupiter together With which he divided domination above the world after overthrow titans. By lot to him got in destiny sea, then how crowned thunderer settled on the Olympus and became to rule mountain heights. Them third the uterine offspring - the terrible Hades (his other name is Pluto) - settled in the "gloomyabysses land" and became lord of the kingdom the dead.

More than one and a half years have passed since the discovery of the eighth planet in the solar system. centuries, but one Neptune year blows only in 2011, since Neptune, distant from sun on the four With half billion kilometers (or thirty astronomical units), commits full cycle per 165 earthly years. By their physical parameters he fewdifferent from Uranus, slightly inferior to it in size (Neptune's diameter is almost 49 530 kilometers), but perceptibly surpassing on mass (17 masses our planets) what explained his greater middle density (about 1.64 g/cm3). From sun Neptune receives in 900once less heat, how Earth. However in difference from calm uranium intensitythermal radiation bowels eighth planets nearly triple exceeds inflow solarenergy from outside. This phenomenon is associated with the decay of heavy radionuclides in the core of the planet.because of huge remoteness Neptune the study his surfaces associated co significant difficulties. However, the need for inventions is cunning. Taking advantagetaunique mutual arrangement of the Earth and the giant planets, space probe Voyager 2 managed slip in 1989 year on the distance 5000 kilometers from Neptune having managed make out some details his cloudy fur coats. AT southern hemisphereplanet discovered A large dark spot the size of Earth, rapidly drifting in westbound at a speed of 325 meters per second. Winds blowing in the atmosphere Neptune is also not a pound of raisins: their speed reaches 400-700 meters per second. Earthly hurricanes tearing roofs off houses and overturning trains, on thisthe background is nothing more than a gentle sea breeze. The planet has a magnetic field, twice inferior in power to the magnetic field of Uranus, as well as a system of rings, some of which present open education like arches.

Like all other gas giants, Neptune is a hydrogen-helium world, and on the share of helium accounts for no more than 15%, and methane is even less - about 1%. Specialists suppose what under cloudy layer lies extensive water ocean, saturated ions various chemical elements.

AT. G. Surdin, one from authors work "Astronomy: century XXI", writes:

...

Significant amounts of methane appear to be stored deeper in the icy mantle. planets. Even at a temperature of thousands of degrees at a pressure of 1 Mbar (one million bar, i.e., a million times more than on the surface of the Earth. – *L. Sh.)* mixture of water, methane and ammonia can form solid ice. To the share of the hot ice mantle, probably accounts for 70% of the mass of the entire planet. About 25% of the mass of Neptune should, according to calculations, belong to the core, consisting of oxides of silicon, magnesium, iron and its compounds, and also rocks. A model of the internal structure of the planet shows that the pressure in itscenter about 7 Mbar, and temperature - around 7000 TO.

Neptune has 13 moons, but the largest one is the most notable. - Triton, having a diameter of 2705 kilometers. Revolving around the mother planet on the distance 355 thousand kilometers (about such same distance separates moon from Earth), he the only one from all satellites Neptune moving on orbit in reverse direction. The surface temperature of Triton does not exceed 38 degrees Kelvin (minus 23 degrees Celsius) and is a fissured plain resembling a melon peel. It is assumed that under the ice shell about 200 kilometers thick lies water ocean 150 km depths, saturated ammonia methane and salts.

However the most big mystery Triton - this is his volcanic activity. Specialists even had to come up with a special term - cryovolcanism volcanism at low temperatures, because no one could have imagined that through frozen worlds on the backyard solar systems may have though some volcanic activity. Imagine yourself geyser, hacking nitric ice on the surface of the planet and taking off to a height of up to 8 kilometers. In this case, the column thickness also very sickly - from 20 meters to 2 kilometers. Jet soaring in the sky dispels winds (at Triton there is sparse atmosphere, consisting from nitrogen, a small amount of methane and hydrogen) and turns into plumes stretching for 150 kilometers.

Triton on 70 % complicated from silicates and on thirty % iso ice, in whose composition are included nitrogen, carbon monoxide and methane. Cryovolcanism has not yet received a clear explanation, but some scientists believe what he maybe to be tied With tidal warming up surfaces planets, a also With penetration solar radiation through translucent upper layers ice.

By comparison With Triton, which the only few less moon, Nereid, having some miserable 340 kilometers across, looks like a perfect crumb. However less this is third on size satellite Neptune before Total interesting topics what draws around maternal planets on extremely elongated orbit With eccentricity near 0.75. Such orbits entirely and beside meet at comets whicheither they approach the Sun, melting in the flames of its chromosphere, or they fly away into darkness and colddistant outskirts solar system.

Nine or ten?

– Tell, gogi, How many will be four times two?
– Seven, teacher.
– Somewhere So, gogi, somewhere So… Seven, eight…

Joke

The ninth planet is spinning in such an utter distance that it is up to beginning of the 20th century was decidedly impossible. Even a beam of light passing through distance from the earth to the sun in just eight minutes, it takes five and a half hours to creep in half to Pluto. Pluto was recently discovered 1930 year, and With moment his discoveries passed few more three With half Pluto's months, for a complete revolution around the Sun, this small and very cold the planet makes almost 246 Earth years. The honor of opening the ninth and smallest planets solar systems belongs American astronomer Clyde Tombo, who at that time was barely 24 years old. However, the fate of Pluto somehow did not immediately wondered. poor guy then bouncers With disgrace from members planetary family, then again accepted back under thunder applause. This stupid leapfrog continued enough for a long time, bye in august 2006 of the year on the General assembly International Astronomical Union in Prague noisy delegates by a majority of votes finally deprived the long-suffering Pluto of the honorary status of a classical planet and did not place his together co satellite Charon in group So called transneptunian objects (TNO). The main reasons for such outrageous discrimination were the small size the ninth planet and some features of its orbit. Pluto is the smallest planet solar systems (total 2300 kilometers in diameter, that is one and a half times less Moon), however, its surface area (17.9 million km2) is quite comparable with the territory Russia.

Pluto, half-brother of Zeus-Jupiter and Poseidon-Neptune, was the ruler the realms of the dead, and Saturn and Uranus were his father and grandfather, so he is wonderful fit into the family of the most distant planets in the solar system. The ancient Greeks considered it rare rich man for to him belonged not only souls dead, but and countless treasures hidden in the depths of the earth. The lord of ancient Erebus had another name - Hades, or Hades, which translates as "formless", "invisible", "terrible". When in 1978 American astronomer James Christie discovered Pluto's natural satellite he was almost immediately christened Charon after the mythical boatman from the realm of the dead. This gloomy and unfriendly old man, dressed in shabby rags, transported the dead along waters underground rivers, which in Aide It was full-full: stormy Styx, fiery Phlegeton, Lethe - the river of oblivion and impenetrable black Cocytus. Alas, everything in the world has my price, a because labored Charon by no means not is free. Remember Brodsky, reader?

In vain the sullen Charon searches for the drachma in your mouth, in vain someone trumpets upstairs in my tune drawn out.

*I send you a nameless farewell With shores
unknown what. Yes you and not important.*

True, Iosif Alexandrovich got a little excited, shamelessly raising the payment for travel. The deceased really put money under the tongue during the funeral rite, however, it was not a full-weight drachma, but an obol - a small silver or copper coin dignity in one sixth her part.

The good world will not be named after the god of death. Compared to Earth, Pluto gets one and a half thousand times less solar heat, therefore, on its surface always reigns icy cold - from minus 220 before minus 240 degrees Celsius. At such low temperatures, even nitrogen freezes, forming large transparent crystals up to several centimeters across. Ordinary water ice can also be found on Pluto, however, in small quantities. Frozen carbon monoxide is found in some areas carbon. A traveler who sets foot on the surface of the ninth planet will see a landscape of stunning beauty, an amazing world of perfect geometric shapes like icy halls Snowy queens from fairy tales Hans christian Andersen. Like boy Kayu, he even maybe to attempt fold word "eternity" from transparent crystals, for where, how not on Pluto, you can in full least feel her regal indifference? jet black sky above head in typhoid rashes stars, conglomeration century ice under feet and huge Charon, still hanging in zenith, how reminder about vanity of all things.

Pluto explored from hands out poorly, because what on the today's day this is the only planet solar systems, before which bye more not got neither one space probe. The flight to Pluto is a very difficult technical task, since six billion kilometers separating the ninth planet from the Sun, present a maximum requirements and to problem radio communications With automatic station, and to elements her power supply. Standard solar batteries on the such huge distance completely useless. Nevertheless, in January 2006, the American apparatus New Horizons", which should meet with the lord of the cold worlds in July 2015. If everything goes well, the space probe will continue flying, everything farther away from the sun. Its new target will be Kuiper belt objects - an amorphous cloud through frozen icy boulders, lying per the orbit of Pluto.

AT 1988 year at ninth planets was discovered very sparse atmosphere, presumably consisting from nitrogen, methane, argon and not she. Pressure this nearly weightless haze is completely negligible, which, however, does not interfere with the flow of chemical reactions. Under influence sunny wind atoms nitrogen, carbon, hydrogen and oxygen interact between yourself generating complex organic connections. settling down on the surface planets, they stain her in yellowish pink color. But most a remarkable feature of Pluto's atmosphere is its seasonal metamorphoses associated with change times of the year. By measure approximation to sun temperature starts grow, what leads to the evaporation of nitrogen ice and the "swelling" of the atmosphere. But if Pluto leaves from sun away (his orbit represents yourself strongly elongated ellipse), how the temperature immediately drops, and the gases condense again and fall to the surface planets in form crystals nitrogen ice. Coming seasonal glacial period, and the atmosphere disappears for a long time without a trace. So Pluto is the only a planet in the solar system whose atmosphere is periodically born and dies, as in comets during their movements around the sun.

The parameters of Pluto's orbit also deserve attention. At the time of its opening, located far enough from the Sun, rightfully occupying the place of the ninth planet. But because it orbit It has very significant eccentricity (0.25, then there is noticeably greater than even that of Mercury), the distance to Pluto from the Sun during its year is changing nearly in two times - from 29.6 a. e. in perihelion before 48.8 a. e. in aphelia. So the way

Pluto is sometimes closer to the Sun than Neptune. through the nearest point Pluto passed its orbit in September 1989 and now continues to move away aphelion (the point of maximum distance from the Sun), which will reach only in 2112, and the first complete revolution around the Sun after its discovery will be completed only by 2176. In addition Pluto's orbit is strongly inclined towards plane ecliptic (17 degrees, on 10 degrees more than Mercury), which is also atypical for most planets in the solarsystems.

Axial rotation ninth planets too It has their peculiarities. Corner between Pluto's equatorial plane and its orbital plane is 32 degrees, so when moving in orbit, it rolls from side to side, like a bun. In this sense, he a little recalls Uranus, although at the last one how we remember axial mood more more: the seventh planet actually lies on its side. Full rotation around Pluto's axis completes in 6.4 Earth days, and its satellite Charon wraps around the mother planets in accuracy per then same most time. Except Togo, orbit Charon lies in equatorial plane of Pluto, so it is visible only from one hemisphere and never not hiding per horizon. BUT because the distance between Pluto and Charon not exceeds 19 400 kilometers, With surfaces Pluto his satellite looks very impressively: his visible diameter in seven once more diameter Moon on the earthly firmament.

I must say that Pluto and Charon are a completely unique tandem among others planets solar systems. They are very close on sizes (2300 and 1200 kilometers respectively) and located on the small distance friend from friend. The ratio of their masses is also unprecedentedly high, since Pluto is only eight times heavier than Charon. For comparison: the Moon, which is traditionally considered very a large satellite, 81 times lighter than the Earth, and located much further. Similar the mass ratios of other planets of the solar system and their satellites give incomparably smaller quantities. Let's say satellites Jupiter (not speaking already about satellites Mars) inferior to it in mass by several thousand times. On the other hand, Pluto and Charon are palpably differ in the average density parameter, which allows us to think about their independent origin. Therefore, most astronomers believe that Pluto and Charon are a doubledwarf planet.

Aggregate all these circumstances - extremely elongated orbit ninth planet, strongly inclined to the ecliptic, its very small diameter and mass, the presence extremely non-standard satellite - in the end they prompted the experts decisively and irrevocably banish Pluto from the number of planets in the solar system and place it on the list objects belts Kuiper (OPK).

The reader has already met so many times on the pages of this book with trans-Neptunian objects (or Kuiper belt objects, which is practically the same thing), that the time has come talk about the distant environs of the solar system in more detail. If some the interstellar wanderer looked at the solar system from the side, he would see that it surrounded spherical cloud protoplanetary tel, swarm stone and icy boulders relatively small sizes. According to some estimates, there are several billion, and the total mass of these celestial bodies is comparable to the mass of Jupiter. This spherical shell, remote on the 20–50 thousand astronomical units from sun, named the Oort cloud in honor of its discoverer, the Dutch astronomer Jan Hendrik Oort. Recall that one astronomical unit (1 AU) is the average distance from Earth to the Sun, which is about 150 million kilometers. Thus the cloud Horta is monstrously far away - 20-50 thousand times farther from the Sun than the Earth. Even Pluto is a thousand times closer, since the aphelion of its orbit lies "only" in 50 astronomical units from our luminary. Such distances no longer make sense measure in kilometers, because from the abundance of zeros it starts to ripple in the eyes. So that you reader, could any visually introduce yourself these open spaces, enough to tell, what central part clouds Oort lies in half light of the year from earthly observer. Proxima Centauri, our closest star, is only eight once farther.

Heavenly body, constituents cloud Oorta, slowly revolve around sun, making a complete revolution in several million years. Astronomers believe that from there, With distant periphery solar systems, come So called long-term comets, which are moving on extremely elongated orbits With perihelion below the orbit of Mercury. In this case, the point of their maximum removal is lost in utter distance - in thousands or even tens of thousands of astronomical units from the Sun. Finally, the orbits of the planets lie approximately in the same plane (the plane of the ecliptic), and comets are flying how God on the soul put - under most bizarre corners, from what, actually, and was concluded about spherical form clouds Oort.

But what force pushes the ice fragments from their calm orbits, forcing them to change nearly circular trajectory on the elliptical? Before recent time it was thought what anomalies in the motion of some objects of the Oort cloud is introduced by the total gravitational the impact of almost all the stars of the Milky Way, since long-period comets evenly distributed on firmament. However several years back American astronomer John Matese came up with a sensational hypothesis. Having carefully analyzed trajectories 82's most Good studied long-term comets he came to the conclusion that a distinct selectivity is found in the distribution of their trajectories. About third these comets comes predominantly With one hand, that's why talk about uniform distribution is not necessary. In addition, they all have atypical orbits - too short compared to the orbits of other comets. According to Matese, the reason similar anomalous behavior is not total gravity stars, a influence some massive body - tenth planets solar systems, which pushes out comets from the Oort cloud towards the Sun. According to his calculations, this planet in several times heavier than Jupiter and hides in the very core of the cloud, at a distance about 25 thousand astronomical units (about 0.4 light years), making a complete turnover around the sun per 4–5 million years.

In addition, the orbit of the hypothetical planet is likely to be strongly inclined to plane ecliptic, a herself she is revolves retrograde then there is in direction, directly opposite movement majority planets solar systems. Orbit With such parameters should be unstable, so the planet "X" of John Matese is not native, but came: she is not could form inside gas-dust disk, which the four With half a billion years ago gave birth to the eight classical planets - from Mercury to Neptune inclusive. Consequently, "wrong" tenth planet initially represented yourself homeless wanderer, wandering in interstellar space, and only relatively recently was she blue-coloured and adopted, when she happened to be in surroundings Sun.

However, it is not yet necessary to talk seriously about the tenth planet in the Oort cloud, because it's real no one was watching - it exists exclusively "on the tip pen" John Matese. BUT here in belt Kuiper which the starts nearly straightaway same per orbits of Neptune and Pluto, many planets have recently been discovered. American astronomer Gerard Kuiper in the 50s of the last century put forward the hypothesis that on out in the back of the solar system, there is a vast number two asteroid belt (as opposed to from Good famous belts asteroids between orbits Mars and Jupiter), which the stretches for billions of kilometers and gradually disappears, leaving between themselves and cloud Oort imposing empty gap. Long time hypothesis American remained nothing more than an elegant game of the mind, until in the early 90s of the last century, several icy debris were not found in the orbit of Pluto. Since then existence the Kuiper belt has become an indisputable fact, and the list of trans-Neptunian objects from year to year steadily is replenished new representatives.

If a cloud Oort liken distant Moscow suburbs then belt Kuiper lying on the distance from 30 to 100 astronomical units from the Sun, will be near Moscow. By estimated specialists, he maybe count hundreds thousand or even millions icy and boulders of various sizes. Tandem Pluto - Charon also fell into the number objects of the Kuiper belt, having lost the status of a classical planet, which we already wrote. Cause to that become small dimensions ninth planets (diameter Pluto just 2300 kilometers, in one and a half times less, how at Moon) and peculiarities her orbits (expressed eccentricity and perceptible incline to plane ecliptic).

serious Pluto's problems started in 2003 year, when Group American astronomers in chapter With Michael Brown discovered in belt Kuiper enough bright an object that received the catalog number 2003UB313. In 2005, it was possible to calculate it orbit and calculate the size of the new planet. It turned out that she was moving extremely elongated orbit and today located in point maximum removal from sun, on the a distance of 97 astronomical units. But when it reaches perihelion, it will be located three times closer - almost the same distance from the Sun as Pluto. Truth, this is happen not soon, because Xena (precisely So named my planet Brown, in honour heroines famous series about warrior woman) commits full turnover around Sun for 650 years. Brown and his team estimate that Xena's diameter should be about 3000 kilometers, which immediately put Pluto in an awkward position, because its diameter significantly less. In addition, Brown's team discovered two more bright Kuiper belt objects.on the distance 51 astronomical units from sun, only a little inferior in sizes ninth planet (about 70 % her diameter).

BUT when It revealed, what diameter xena, Maybe, determined wrong, a true her dimensions may in two With superfluous times exceed diameter Pluto passions and at all flared up not on the joke. FROM which such, one asks by the way we must count his ninth and latest planet solar systems, if a lot of farther around sun a much more impressive celestial body turns? Isn't it easier to ruthlessly clean unlucky Pluto from a friendly planetary family, reclassifying it into a belt object Kuiper? Especially when you consider that Xena found a satellite, named Gabrielle in honor of true girlfriends brave warriors. AT brackets note what subsequently Xena renamed Eridu - the ancient Greek goddess of enmity and discord, cutting her diameter before 2400 kilometers. Tem not less he all equals more Pluto diameter whom is 2300 km. Gabriela too crossed out from saints - today she is called Dysnomia. By the way, it was Eris who quarreled with Aphrodite, Athena and Hera, throwing table the famous apple of discord with the inscription "Most Beautiful", which led to the Trojan war. Good that at Greeks there were so many gods...

In early 2004, the American Spitzer space telescope found in the belt Kuiper more one planet which now located in 13 billion kilometers from Sun, that is, twice as far away as Pluto. Like Xena-Eris, she moves on godlessly stretched ellipse, making one revolution around the Sun in 10,500 years. Its aphelion (point maximum distance) should lie 130 billion kilometers from our luminary, which is about 900 astronomical units, so the dimensions of the Kuiper belt should be, likely to increase by at least an order of magnitude. The new planet was named Sedna honour eskimo goddesses ocean and mistresses maritime animals, a her diameter estimated at 1800 kilometers. Among other finds of the "zero" years, there are several more notable objects: dwarf planets 2003EL61 and 2003FY9, almost as good as Pluto in sizes, and Quaoar With across near 1300 kilometers (Quaoar - this is creator deity at Indians tribe Tongva). First from these planets It has form ellipsoid rotation and travels in escorted two satellites.

The Kuiper belt has given astronomers a lot of mysteries. For example, it turned out that he thins out gradually, as its discoverer believed, and abruptly and unexpectedly breaks off at some - very big - distance from Sun. By opinion specialists, similar
"beheading" explained explosion supernova stars nearby from our luminaries, in

as a result of which the entire marginal part of the gas-dust cloud, which served as material for the formation of the planets of the solar system, turned out to be completely swept. Initial the idea of the Kuiper belt as a flat disk of protoplanetary bodies (as opposed to spherical clouds Oort) too, apparently should recognize erroneous. Let's say the orbit of Xena-Eris is not only strongly elongated, but also inclined to the plane of the ecliptic under angle of 44 degrees, and the angle of inclination of the orbits of two other Kuiper belt objects discovered group brown, is 28 degrees. BUT if recall, what is the orbit Pluto too lies outside the plane of orbits of all the other planets of the solar system (though Pluto this corner less - Total 17 degrees), then already only on this parameter should exclude from the list classical planets.

Thus, the orbits of almost all Kuiper belt objects are inclined to the plane of the ecliptic is completely arbitrary, which strongly contradicts the current prevailing theory of the formation of planets in the solar system. Judging by the orthodox scenario, planets were born from a flat disk of gas and dust that surrounded the maturing in his center star - future Sun. However latest observational data irrefutably testify what belt Kuiper - no not belt and his it is forbidden treat it as a flat disk. Most likely, it is a spherical a formation resembling the much more distant Oort cloud. Then our solar system, if look on the her from the outside will be similar on the matryoshka or bulb: one big sphere (cloud Oort), inside her sphere slightly less (belt Kuiper) and, finally, Sun and eight planets lying practically in one planes.

The old theory of the origin of the planets does not give such a picture, therefore, in recent years some astronomers began to actively develop a fundamentally different scenario, which received the name of the oligarch. Within this version, the main role is assigned to the so-called oligarch planets, which, by the power of their gravity, significantly influenced the behavior other celestial bodies. After the birth of the Sun, classical planets and asteroid belts the process of formation of the solar system was by no means completed, but continued to gain turns. Asteroids grew rapidly and after crossing a certain limit began to vigorously attract to yourself other body, turning into in large planets. Sergey Ilyin in article
"Stormy biography tenth planet" published in June room magazine "Knowledge
– strength" per 2006 year, detail sets out essence oligarchic script.

...

According to the authors of this new theory, the same process simultaneously occurred on the outskirts of the solar system, in the Kuiper belt. As a result, as calculations show computer simulations, inside solar systems must It was form 20–30 objects the size of Mars, and on the outskirts - about the same number of objects the size of the Earth. With such a number, they should have been close enough, and this with the need caused distortion them orbits friend friend. Traffic "oligarchs" became chaotic they "thrown out" friend friend With sustainable orbits, located in plane ecliptic. Part from them at this generally was leaving from solar systems in interstellar space, becoming "homeless" planets, "planetary" other, the rest acquired orbits inclined at the most "wild" angles to the plane ecliptic, and thus in their totality created a spherical cloud with a diameter of 1000 astronomical units or more. In this cloud, therefore, must to this day day exist not only "small planets" type Pluto or 2003UB313, but and some of the survivors "primary oligarchs". Proponents of such a scenario hope what created now telescopes, destined for goals timely warnings Earth about asteroid danger, allow parallel produce systematic search for such "oligarchs" and find "the tenth, eleventh, twelfth and So Further" planets with land or even more.

Well what and, let's live - we'll see...

BUT how this is the case a business With planets near others stars? After all if our Sun, representing yourself ordinary yellow star spectral class g, managed acquire an impressive family of eight classical planets and tens of thousands offsuit asteroids and dwarf planets, it is logical to assume that other stars can also have their own planets. And since the main haven of life in The universe is precisely the planets (in any case, most tend to think so biologists), the search for extrasolar planets is of particular relevance. Indeed, the conclusion indispensable "binding" life to surfaces planets made on the basis our very meager experience (life is known to us in a single earthly version), but fortune-telling coffee shop thicker more less fruitfully. Of course, quite probably, what life maybe be born even in interstellar environment (in his time English astrophysicist Fred hoylewrote on the this topic fantastic novel under name "Black cloud"), but such a hypothesis would be even more speculative. With the planets, it's somehow clearer - to that example our own Existence. That's why if we want know, how much life is common in the Universe, you must first deal with planetary systems inothers stars.

Until recently, many scientists believed that planets are a very rare occurrence in space. Such sight With evidence flowed out from theories origin planets English astronomer Jeans. According to this once popular theory, planets The solar system was formed from the tongue of the solar substance, which was snatched gravitational forces of a massive star passing by the Sun. jet of matter, splashed out into space, had a spindle shape - with a thickening in the central parts and relatively thin ends. Therefore, the closest planets to the Sun groups and the most distant ones like Pluto and other Kuiper belt objects are small in size. sizes and mass, a in center solar systems settled gas giants. BUT since the approach of stars is not only an accidental event, but also extremely rare (in any case, in the outskirts of the Milky Way, where our Sun is), the birth of planetary systems committed very infrequently. Truth, today theory Jeans represents in significant measure historical interest, So how on the shift her came different scenario: practically simultaneous occurrence planets and sun from rotating gas and dust cloud. Be that as it may, theories remain theories, and we wish know for sure are there planetary systems at others stars.

Of course direct optical observation planets near others stars impossible even today and is unlikely to be possible in the foreseeable future. And although scientific and technical progress hurries forward by leaps and bounds, there are prohibitions on fundamental character. planets, how known present yourself heavenly body, which shine by the reflected light of their sun, so their brilliance against the backdrop of the radiance of the mother star practically indistinguishable. To see a delicate spark against the background of a blazing fire so far so far no one has been able to. Possibly at the center of the Milky Way, where the stars collideinto close flocks, visual tracking of the planets is not particularly difficult, but on periphery our galaxies fixation planets at neighboring stars turns around nearly an unsolvable task. Spiral arms of the Milky Way, one of which vegetates our Sun, distant from center galaxies on the 26 thousand light years, not may boast of high density stellar population. it by no means not Holland, not Belgium and not the Ganges valley, where people sit on each other's heads, but rather Yakutia or Chukotka. There is a lot of free space in our galactic latitudes. I'll remind you reader that even the nearest stars lie unimaginably far away: the distance to Proxima Centauri (by the way, "proxima" in Latin means "nearest") is 4.3 light of the year, famous "flying" star Barnard lags behind from sun on the 6 light years, a to Sirius - most bright star our sky - nearly 9 light years.

If you take a cube with a side of 10 light years, then at best they will fit in it two or three stars. BUT here in ordinary ball congestion, lying not far from center galaxies (in composition milky Ways such clusters near 200) on the 100 cubic light years account for several hundreds stars. Density stellar population there in several thousand times higher, and the night sky in those parts must be unusually bright. So, emphasize more once: direct optical observation out-solar planets (or exoplanets, how them become call today) not seems possible.

But if exoplanet it is forbidden discover directly, then, to be maybe, in disposal contemporary astronomy there are indirect methods them detection? AT Currently, several such methods have been proposed - the astrometric method, the method radiation speeds, observation transits and some other. I not I will become go into in technical details and break down each of these approaches, but I will only note that majority contemporary methods detection exoplanets based on the accounting gravity disturbances in movement stars. A business in volume, what any massive body(for example, a planet), revolving around a star, acts on it with the force of its gravity. In this case, the planet, as it were, slightly pulls the star towards itself, and since due to movement along orbit she is periodically turns out on various sides from luminaries, then and star periodically is shifting in different directions under action gravity planets. Others words if planet moving on orbit around maternal stars, then and star, in my turn, not remains motionless, a describes tiny circle in space under influence forces gravity his natural satellite. So the way both bodies actually revolve around a common center of mass, which astronomers called barycenter.

Of course weight planets negligible small on comparison With weight stars, that's why scope her hesitation very small. Let's say Sun under impact attraction Jupiter (and this is the most massive planet) oscillates about the center of mass of the solar systems at a speed of only 12.5 meters per second. For Earth or Venus, this value is still less and is about 0.1 meters per second. We can say that the sun is a little swaying at movement planets on their orbits a barycenter solar systemslies, so the way inside our luminaries. Before most recent time sensitivity equipment, available in disposal astronomers was clearly insufficient to detect light celestial bodies around other stars. Although such attempts repeatedly were made all they were on the limit experimental accuracy and were subjected reasonable doubt.

The situation changed only in the early 1990s, when spectrometers of a new generation, which made it possible to measure radial velocities much more accurately stars. What such radial speed? If a at stars available satellite (other star or planet), then at movement around barycenter radial speed stars (speed her approaching or moving away from the observer along the line of sight) will experience fluctuations with period, equal period circulation stars around center wt. Sensitivity equipment in end XX century increased, on extreme least on the order, So what became possible find extrasolar planets, comparable on mass With Jupiter.

In addition to the astrometric method and the radial velocity method, there is another way detection exoplanets - So called observation transits. If a to catchplanet at the moment of its passage through the disk of a star, it is possible not only to calculate its mass, but and define dimensions (volume), a Consequently - calculate density. Of course it is impossible to distinguish a dark circle on the dotted disk of a star (even with the most powerful telescope stars look like dimensionless points), but to measure a small decrease in flux Sveta from stars quite Maybe. To unfortunately method observations transits requires fulfillment special conditions: planet, her star and terrestrial observer must be located in one plane (in plane Keplerian orbit, how they say astronomers). Such luck falls out relatively rarely, that's why cases observations transits can literally be counted on the fingers. Nevertheless, the game is worth the candle, because only with the help of this method it is possible to study a number of important characteristics of exoplanets, measure them radius and even research properties them atmospheres.

The first success fell out on the share Swiss astronomers M. Major and D. Quelotsa, which lucky discover planet near sun-like stars, designated in directory how 51st in constellation Pegasus (51 Peg). it significant event happened in 1994, but the characteristics of the first exoplanet were so unexpected what scientists decided detain publication, to how should recheck their results. By 1995, all doubts were gone, and the discovery hatched. New planet at 51 Pegasus was amazing. Its mass was approximately equal to the mass of Jupiter, and the distance from maternal stars was Total 0.05 astronomical units, then there is in twenty once less than from the Earth to the Sun (and even almost 8 times less than from the Sun to Mercury). Planet committed full turnover around stars per 4.2 days - such is was duration her of the year. because of closeness to luminary temperature her surfaces exceeded 1000 degrees by Kelvin.

To tell, what scientific world was overthrown in condition shock - nothing not to tell. planetary system 51 Pegasus turned out absolutely dissimilar on the solar system. In the fall of 1995, Major and Quelotz reported their discovery at a conference in Italy, and planets agreed call on name stars With adding letters "b" for first found planets, "With" - for second and So Further. At first astronomers amused myself hope what Swiss managed stumble on the some anomaly unprecedented rarity in world planets, but subsequent finds forced take a look on the things differently. Another exoplanet had a mass four times greater than that of Jupiter, and the period its revolution around the parent star (that is, the year) turned out to be even shorter - 3.3 days. Subsequently, planets of this type began to be called "hot Jupiters". True, in In 1996, the American astronomers D. Marcy and P. Butler seem to have managed to discover planetary system, partly reminiscent solar, at stars upsilon Andromedae (? And), but more attentive analysis showed what resemblance this is apparent. AT system
?And three very weighty planets are circling around the parent star, and the mass the nearest of them is slightly less than the mass of Jupiter, and the other two are heavier than our gas giant in two and four times respectively. First (most easy) planet - typical
"hot Jupiter" with an orbit radius of 0.06 AU. e., but the other two lie on quite decent distances - 0.9 and 2.5 a. e. However, the orbits of these distant exoplanets have nothing in common With orbits planets solar systems, because the possess very significant eccentricity. Unfortunately, it's a bummer again. The list of extrasolar planets continued steadily replenish, and to middle Martha 2007 of the year there were already 182 stars, burdened by planets. And since in some systems it was possible to find several planets, them general amount outnumbered 200.

Thus, today astronomers have, albeit limited, but However, there are enough statistics to support the assertion that Approximately 4% of stars close to the Sun in terms of spectral properties have planetary systems or single planets. Slightly hotter and slightly colder stars classes F and K (recall that our Sun belongs to class G) planets were found completely few. Of course, this does not mean that hot white and blue stars do not have planets. in reality; it's just that the radial velocity method is not universal and does not work well if star has a restless photosphere.

But the main problem is that almost all newly discovered exoplanets or planetary families show a striking difference from the solar systems and her planets. Only in single cases managed discover planets, circulating on circular or nearly circular orbits on the sufficient removal from maternal stars. All others or are spinning how crazy, back to back to his the sun warming up to hundreds and thousands of degrees (and we are talking about gas giants the size of Jupiter, a then and more), or are on the sharp eccentric orbits more resembling the orbits of comets. What would you say about a planet several times larger than on mass Jupiter, which then approaching to maternal star nearly back to back then flies beyond the orbit of Neptune? Meanwhile, this is exactly how the planetary families strangers suns.

Recently, astronomers have been talking about "very hot Jupiters." One such planet, in one and a half times exceeding Jupiter on mass, was relatively recently discovered at stars sunny type. She is located on the distance 3.3 million kilometers (0.02 AU) from the parent star (the average distance of Mercury from the Sun is 58 million kilometers) and revolves around it in a record short time - 1.2 days. From the surface of this unique planet, the mother star looks unimaginable. a huge ball bursting with sizzling fire (50 times larger in diameter than the Sun on the earthly sky).

Unusual planetary families others stars resolutely contradict generally accepted theory of the formation of planetary systems, according to which the Sun and planets were born from gas-dust disk practically simultaneously. All planets solarsystems fall into two large groups: relatively small solid balls with high density, folded rocky breeds, and gas giants, whose average density few is different from density water. Difference between big and small planets explained topics what gas giants were born in central parts protostellar cloud by gradually accumulating huge masses of gas on the primary icy nucleus, a small planets formed on the near and distant periphery gas-dust disk, where substances It was very not much. Education planets terrestrial groups conceived how result multiple clashes and mergers So called planetazimal (planetary embryos) With subsequent them warming up per check radioactive elements, settled in nuclei solid planets. Because the primary gas-dust cloud had form rotating around vertical axes disk With thickening in the center, the orbits of all the planets should be almost regular circles and lie in the same plane. At least that's what the generally accepted theory says. planet formation.

Meanwhile, exoplanets and exoplanet families stubbornly refuse to fit into this idyllic picture, so astrophysicists and planetary scientists have to look for other explanations. And if unusual properties first extrasolar planets at first regarded as some kind of anomaly, then new discoveries encourage us to think about what anomaly quicker Total, should count our solar system. To explain phenomenon of "hot Jupiters", a migration mechanism was proposed, which is slow sliding planets With high orbits, where they originally formed, on the orbits low, circumstellar. That circumstance, what they neither in who case not could born in close proximity to the mother star, where they are to this day, majority planetary scientists doubt not calls. Additional argument in benefit
"distant" birth "hot Jupiters" are discovered astronomers gas-dust clouds in the stage of planet formation. The vast area around the star is always cleanly swept, free from dust and gas, because the density of stellar radiation here so high that it completely sweeps all the garbage to the periphery. Therefore, the material which low-orbit "hot Jupiters" are formed, can only be located on distance not less five astronomical units from parental stars. By all visibility, mechanism migration turns on very early, a developments develop very rapidly: barely having time be born planets start slide on gently sloping spirals to his the sun bye tidal interactions stars and planets not stabilize orbit
"hot Jupiter" back to back to star. However, quite available and another scenario: gravity maternal stars constantly slows down planet bye that not collapse on

tapering spirals on the his sun and not burn down in his bowels.

Squeezed close to the parent star, gas giants are so an ordinary phenomenon that can only shrug. The solar system phenomenon finds intelligible explanations. Doctor physical and mathematical Sciences L. xanfomality, employee Institute space research RAS, writes about this next way:
"Extrasolar planets offer theorists so many questions that it fits the whole theory education planets write again. BUT naive question: why migration No in our solar system? - them better not set". Tem more not costs to ask specialistsabout others physical parameters exoplanets. Taking per point reference solar system, we have the right to assume that the average density of gas giants near alien suns (hot they or cold - fundamental values not It has) must fit into familiar values, a little different from density water. However, not here and there It was! Medium density massive exoplanets "floats" in very wide within
– from half the density of Jupiter to several densities of Saturn. For example, one of such planets, significantly inferior to Jupiter in diameter, thoroughly surpasses it in mass, from what should suppose what she is has weighty core from heavy elements, on the which account for before 0.7 masses new exoplanets. Gas giants in The solar system cannot to boast of such a dense core, so in standard theories origin planets this fact not finds intelligible explanations.

The phenomenon of "hot Jupiters" astrophysicists have explained in half, but remain more "cold jupiters", entirely and beside describing around maternal stars so stretched ellipses, which more stuck long-term cometsfrom time to time flying away to nowhere. True, computer simulation seems to be helped shed light on the evolution planetary systems upsilon Andromedae ("hot Jupiter" in low orbit and two distant planets with a distinct orbital eccentricity). FROM another hand, models models strife. For example, employees Washington university in Seattle for some reason came to conclusion what majority exoplanets, similar in size with the Earth (just in case for reference: not a single such planet has yet been was observed for them detection lies per outside contemporary astrophysical methods), must to be water worlds. They are shuffled various scenarios planetogenesis,and each time four Earth-like planets appeared on the display, the smallest of which was fivefold less earth, a the most big - in four times more. At computer modeling on the these virtual lands accumulated incredible the amount of water is 300 times more than on the real Earth, so their entire surface must to be covered impressive ocean many kilometers depths.

By the way, what about the search for terrestrial-type planets? Alas, practically nothing, So how sensitivity method radiation speeds allows reliably detect only giant planets (planets near pulsars, which will be discussed below is a rare and happy exception). The smallest of the recently discovered exoplanets revolves around red dwarf - stars spectral class M With temperature surface is 2-3 thousand degrees Kelvin (our Sun has 6 thousand). Presumablyit is solid, that is, it consists of rocks, like the Earth, and its mass is estimated about 7.5 Earth masses (noticeably less than that of Neptune or Uranus). Everything would be nothing however, unfortunately, this is again a planet in low orbit (though due to the relatively small size to call it "Jupiter" somehow the language does not turn). Around your dim sun, it turns in two days (1.94 days) and is at a distance from it three million kilometers - 50 times closer than the Earth from the Sun. And although the red dwarf - not like our hot luminary, it nevertheless warms up the surface of a rapidly flying planets before 200–400 degrees in Celsius. Life earthly type there barely whether possible.

However despair all same not costs, because the statistics extrasolar planets long away not full. Let's say considerable interest represents system stars HD37124 in constellation Taurus where discovered three planets, each from which twice easier Jupiter a the radii of their orbits are 0.5, 1.7 and 3.2 AU. e. And since there is a special tightness in the star system from the constellation of Taurus is not observed, it is quite possible to assume the presence of terrestrial planets there type. The same applies to the star 47 Ursa Major, in which massive planets resembling Saturn and Jupiter, with a very similar parameters orbits. Therefore, in the inner region of this system, the existence of planets earth type.

However, the fact remains that the structure of the orbits of the vast majority of exoplanets even remotely not recalls solar system. back to back squeezed to their hot gas balls to the sun or running away along unimaginably stretched ellipses icy giants not have nothing general With planets solar systems. If ato suggest that in the inner regions of some exoplanetary systems there is room for Earth-like planets, it is difficult to imagine how they can survive, because migration giants to star inevitably will lead to catastrophic intersection orbits.

Even the anatomy of foreign gas giants is fundamentally different. Many of them have massive core from heavy elements, on the which account for before 70% all masses planets. noticeably inferior in size our Jupiter or Saturn, so atypical exoplanets significantly outnumber them in mass. There is nothing like it in the solar system. meets. All these puzzles, together taken, lead to very sad conclusion about uniqueness our planetary systems. planets terrestrial groups apply on sustainable orbits and in principle able to be cradle life. giant planets slowly circling in the distance and do not interfere with anyone; Moreover, there is a point of view according to which they perform important protective function, covering domestic planets from unexpected attacks of dangerous celestial bodies. It comes down to some astrophysicists talk about a peculiar version of the anthropic principle, in accordance with which occurrence life on the Earth closest way related With Jupiter.

Astronomy as a science developed under the sign of increasing decentralization. First we learned, what Earth not is center universe, a represents yourself very a modest celestial body tirelessly scurrying around the sun. Then it turned out that our magnificent luminary, deified, exalted to heaven and giving life to every creatures - an ordinary yellow dwarf of the spectral class G, which are part of the Milky There are darkness on the path. And it is by no means located in the center of the Galaxy, as recklessly believed some astronomers of the XVIII century, and settled on its distant backyards, where there were only a few stars, between two dusty spiral arms. BUT now us they say, what disk milky Ways, this twisted in tight node monstrous blot With across in 100 thousand light years, there is not what other how one from hundreds billion galaxies, scattered on boundless universe.

The thought of the uniqueness of the solar system continues to sit like a splinter, fairly poisoning astronomers life. Xanfomality writes:

...

All large planets solar systems have nearly coplanar (located in one plane) stable orbits With low eccentricity, exclusive them catastrophic convergence. Sunny system - this is system With low entropy (high stability). But it is precisely the high-entropy systems of exoplanets in which only the most massive bodies survive may be the norm. The solar system could be completely different than the one in which we live. Or maybe we live in it exactly because she not similar on the other?

AT conclusion remains to tell, what first exoplanet was discovered not 1994 year, a on the several years before - in 1990 when American astronomer Polish origin Alex Woltzshan (Volchan in another transliteration) sent mine radio telescope to the faint pulsar PSR 1257+12, located at a distance of 1300 light years from Earth. By their physical nature, pulsars are neutron stars. which emit powerful, strictly periodic pulses of electromagnetic radiation. Pulse periodicity everyone has it pulsar strictly individual and usually lies in ranging from 640 pulses per second to one pulse per five seconds. swiftly a rotating neutron star is, in fact, a giant magnet, and along straight, connecting poles this magnet, which the spinning how mad, fly out So called jets - powerful jets red-hot plasma and photons. The brightness variability is explained simply, since the magnetic pole does not have to lie on the axis rotation (the magnetic poles of the Earth also do not coincide with the geographic poles). The outgoing electromagnetic jet describes a cone around the axis of rotation, and we see the pulsar only in those moments when it "looks" directly at the Earth. In a moment he turns away and goes aside, in order to return again after some, strictly fixed time interval.

Because the period pulsars exclusively stable (up to before 10-14 seconds), radiation speed neutron stars can measure With precision before 1 cm/s what completely inaccessible to ordinary stars. Even more precisely, one can define its periodic displacement when revolving around the barycenter, so the pulsar does not have a large labor to detect planets with a mass of the order of the Earth. But since the existence of planets pulsars nobody not could dream even in nightmarish dream, astronomers simply waved on the them hand.

But Alex Woltzschan broke the tradition and did not lose. Analysis of pulsar variations with pulse frequency of 6.2 milliseconds showed that around a neutron star as many as three planets, the masses of which are quite comparable with the mass of the Earth (0.02, 4.3 and 3.9 M „respectively). orbits, on which they are moving nearly circular and constitute 0.20.4 and 0.5 a. e. Periods appeals too acceptable - 25, 66 and 98 days. Problem is in volume, what absolutely unclear, what way these planets could safely survive explosion supernova, for neutron star there is not what other how product of the explosion of an ordinary star at the end of its life. Supernova explosion is monstrous cataclysm, which the must was "to iron" clean up neighborhood stars, So what planets elementary not could survive. Astrophysicists suppose what nearby from exploded supernova once there was another star, the substance of which gradually flowed to the pulsar (a pulsar is a very massive body), and the snot that remained out of work, condensed in planets.

To decide, how much unique Sunny system, need continue Search exoplanets, and in the first place - Earth-like. There is reason to believe that the future the decade should be marked by new discoveries. The French intend to launch space satellite COROT, specially designed to observe transits, and American orbital telescope "Kepler" for four years of work will be able to explore near 100 thousand stars. European space agency planned launch satellite
"Darwin", representing yourself system from six orbital telescopes, which the aims to search for chemical signs of life on other planets. It remains to be hoped that amount early or late will pass in quality.

www.ingramcontent.com/pod-product-compliance
Lightning Source LLC
Chambersburg PA
CBHW060417220526
45465CB00008B/2920